T0301801

ROADS TO INFINITY

ROADS TO INFINITY

THE MATHEMATICS OF TRUTH AND PROOF

JOHN STILLWELL

CRC Press
Taylor & Francis Group
Boca Raton London New York

CRC Press is an imprint of the
Taylor & Francis Group, an **informa** business

AN A K PETERS BOOK

CRC Press
Taylor & Francis Group
6000 Broken Sound Parkway NW, Suite 300
Boca Raton, FL 33487-2742

© 2010 by Taylor & Francis Group, LLC
CRC Press is an imprint of Taylor & Francis Group, an Informa business

No claim to original U.S. Government works

ISBN 13: 978-1-56881-466-7 (hbk)

Visit the Taylor & Francis Web site at
http://www.taylorandfrancis.com

and the CRC Press Web site at
http://www.crcpress.com

Library of Congress Cataloging-in-Publication Data

Stillwell, John.
    Roads to infinity : the mathematics of truth and proof / John Stillwell.
        p. cm.
    Includes bibliographical references and index.
    ISBN 978-1-56881-466-7 (alk. paper)
        1. Set theory. 2. Intinite. 3. Logic, Symbolic and mathematical. I. Title.

QA248.S778 2010
511.3'22–dc22   2010014077

TO ELAINE

CONTENTS

PREFACE

Mathematics and science as we know them are very much the result of trying to grasp infinity with our finite minds. The German mathematician Richard Dedekind said as much in 1854 (see Ewald (1996), pp. 755–756):

> ...as the work of man, science is subject to his arbitrariness and to all the imperfections of his mental powers. There would essentially be no more science for a man gifted with an unbounded understanding—a man for whom the final conclusions, which we attain through a long chain of inferences, would be immediately evident truths.

Dedekind's words reflect the growing awareness of infinity in 19th-century mathematics, as reasoning about infinite processes (calculus) became an indispensable tool of science and engineering. He wrote at the dawn of an era in which infinity and logic were viewed as mathematical concepts for the first time. This led to advances (some of them due to Dedekind himself) that made all previous knowledge of these topics seem almost childishly simple.

Many popular books have been written on the advances in our understanding of infinity sparked by the set theory of Georg Cantor in the 1870s, and incompleteness theorems of Kurt Gödel in the 1930s. However, such books generally dwell on a single aspect of either set theory or logic. I believe it has not been made clear that the results of Cantor and Gödel form a seamless whole. The aim of this book is to explain the whole, in which set theory *interacts* with logic, and both begin to affect mainstream mathematics (the latter quite a recent development, not yet given much space in popular accounts).

In particular, I have taken some pains to tell the story of two neglected figures in the history of logic, Emil Post and Gerhard Gentzen. Post discovered incompleteness before Gödel, though he did not publish his proof until later. However, his proof makes clear (more so than Gödel's did) the origin of incompleteness in Cantor's set theory and its connections with the theory of computation. Gentzen, in the light of Gödel's theorem that the consistency of number theory depends on an assumption from outside number theory, found the *minimum* such assumption—one that grows out of Cantor's theory of ordinal numbers—paving the way for new insights into unprovability in number theory and combinatorics.

This book can be viewed as a sequel to my *Yearning for the Impossible*, the main message of which was that many parts of mathematics demand that we accept some form of infinity. The present book explores the consequences of accepting infinity, which are rich and surprising. There are many levels of infinity, ascending to heights that almost defy belief, yet even the highest levels have "observable" effects at the level of finite objects, such as the natural numbers 1, 2, 3, ... Thus, infinity may be more down-to-earth than certain parts of theoretical physics! In keeping with this claim, I have assumed very little beyond high school mathematics, except a willingness to grapple with alien ideas. If the notation of symbolic logic proves *too* alien, it is possible to skip the notation-heavy parts and still get the gist of the story.

I have tried to ease the reader into the technicalities of set theory and logic by tracing a single thread in each chapter, beginning with a natural mathematical question and following a sequence of historic responses to it. Each response typically leads to new questions, and from them new concepts and theorems emerge. At the end of the chapter there is a longish section called "Historical Background," which attempts to situate the thread in a bigger picture of mathematics and its history. My intention is to present key ideas in closeup first, then to revisit and reinforce them by showing a wider view. But this is not the only way to read the book. Some readers may be impatient to get to the core theorems, and will skip the historical background sections, at least at first reading. Others, in search of a big picture from the beginning, may begin by reading the historical background and then come back to fill in details.

The book has been developing in my unconscious since the 1960s, when I was an undergraduate at the University of Melbourne and a graduate student at MIT. Set theory and logic were then my passions in mathematics, but it took a long time for me to see them in perspective—I apologize to my teachers for the late return on their investment! I particularly want to thank Bruce Craven of Melbourne for indulging my interest in a field outside his specialty, and my MIT teachers Hartley Rogers Jr. and Gerald Sacks for widening my horizons in logic and set theory.

In recent times I have to thank Jeremy Avigad for bringing me up to date, and Cameron Freer for a very detailed criticism of an earlier draft of this book. John Dawson also made very helpful comments. If errors remain, they are entirely my fault. The University of San Francisco and

Monash University provided valuable support and facilities while I was writing and researching.

I also want to thank my friend Abe Shenitzer for his perennial help with proofreading, and my sons Michael and Robert for sharing this onerous task. Finally, I thank my wife Elaine, as always.

John Stillwell
University of San Francisco and Monash University
March 2010

- Chapter 1 -

The Diagonal Argument

Preview

Infinity is the lifeblood of mathematics, because there is no end to even the simplest mathematical objects—the positive integers 1, 2, 3, 4, 5, 6, 7, ... One of the oldest and best arguments about infinity is Euclid's proof that the prime numbers 2, 3, 5, 7, 11, 13, . . . form an infinite sequence. Euclid succeeds despite knowing virtually nothing about the sequence, by showing instead that any *finite* sequence of primes is incomplete. That is, he shows how to find a prime p different from any given primes $p_1, p_2, \ldots, p_n$.

A set like the prime numbers is called *countably infinite* because we can order its members in a list with a first member, second member, third member, and so on. As Euclid showed, the list is infinite, but each member appears at some finite position, and hence gets "counted."

Countably infinite sets have always been with us, and indeed it is hard to grasp infinity in any way other than by counting. But in 1874 the German mathematician Georg Cantor showed that infinity is more complicated than previously thought, by showing that the set of real numbers is *uncountable*. He did this in a way reminiscent of Euclid's proof, but one level higher, by showing that any countably infinite list of real numbers is incomplete.

Cantor's method finds a real number x different from any on a given countable list $x_1, x_2, x_3, \ldots$ by what is now called the *diagonal argument*, for reasons that will become clear below. The diagonal argument (which comes in several variations) is logically the simplest way to prove the existence of uncountable sets. It is the first "road to infinity" of our title, so we devote this chapter to it. A second road—via the *ordinals*—was also discovered by Cantor, and it will be discussed in Chapter 2.

1

1.1   COUNTING AND COUNTABILITY

> If I should ask further how many squares there are, one might reply truly that
> there are as many as the corresponding number of roots, since every square
> has its own root and every root has its own square, while no square has more
> than one root and no root more than one square.
>
> —Galileo Galilei,
> *Dialogues Concerning the Two New Sciences*, First day.

The process of counting 1, 2, 3, 4, ... is the simplest and clearest example
of an *infinite process*. We know that counting never ends, because there
is no last number, and indeed one's first thought is that "infinite" and
"neverending" mean the same thing. Yet, in a sense, the endless counting process *exhausts* the set $\{1, 2, 3, 4, \ldots\}$ of positive integers, because
each positive integer is eventually reached. This distinguishes the set of
positive integers from other sets—such as the set of points on a line—
which seemingly cannot be "exhausted by counting." Thus it may be
enlightening to dwell a little longer on the process of counting, and to
survey some of the infinite sets that can be exhausted by counting their
members.

First, what do we mean by "counting" a set of objects? "Counting"
objects is the same as arranging them in a (possibly infinite) *list*—first
object, second object, third object, and so on—so that each object in the
given set appears on the list, necessarily at some positive integer position. For example, if we "count" the squares by listing them in increasing
order,

$$1, \ 4, \ 9, \ 16, \ 25, \ 36, \ 49, \ 64, \ 81, \ 100, \ 121, \ 144, \ 169, \ 196, \ 225, \ \ldots,$$

then the square 900 appears at position 30 on the list. Listing a set is
mathematically the same as assigning the positive integers in some way
to its members, but it is often easier to visualize the list than to work out
the exact integer assigned to each member.

One of the first interesting things to be noticed about infinite sets
is that *counting a part may be "just as infinite" as counting the whole*. For
example, the set of *positive even numbers* 2, 4, 6, 8, ... is just a part of
the set of positive integers. But the positive even numbers (in increasing
order) form a list that matches the list of positive integers completely,
item by item. Here they are:

1	2	3	4	5	6	7	8	9	10	11	12	13	...
2	4	6	8	10	12	14	16	18	20	22	24	26	...

Thus listing the positive even numbers is a process completely parallel to the process of listing the positive integers. The reason lies in the

item-by-item matching of the two lists, which we call a *one-to-one correspondence*. The function $f(n) = 2n$ encapsulates this correspondence, because it matches each positive integer n with exactly one positive even number $2n$, and each positive even number $2n$ is matched with exactly one positive integer n.

So, to echo the example of Galileo quoted at the beginning of this section: if I should ask how many even numbers there are, one might reply truly that there are as many as the corresponding positive integers. In both Galileo's example, and my more simple-minded one, one sees a one-to-one correspondence between the set of positive integers and a part of itself. This unsettling property is the first characteristic of the world of infinite sets.

COUNTABLY INFINITE SETS

A set whose members can be put in an infinite list—that is, in one-to-one correspondence with the positive integers—is called *countably infinite*. This common property of countably infinite sets was called their *cardinality* by Georg Cantor, who initiated the general study of sets in the 1870s. In the case of finite sets, two sets have the same cardinality if and only if they have the same number of elements. So the concept of cardinality is essentially the *same* as the concept of number for finite sets.

For countably infinite sets, the common cardinality can also be regarded as the "number" of elements. This "number" was called a *transfinite number* and denoted $\aleph_0$ ("aleph zero" or "aleph nought") by Cantor. One can say, for instance, that there are $\aleph_0$ positive integers. However, one has to bear in mind that $\aleph_0$ is more elastic than an ordinary number. The sets $\{1, 2, 3, 4, \ldots\}$ and $\{2, 4, 6, 8, \ldots\}$ both have cardinality $\aleph_0$, even though the second set is a strict subset of the first. So one can also say that there are $\aleph_0$ even numbers.

Moreover, the cardinality $\aleph_0$ stretches to cover sets that at first glance seem much larger than the set $\{1, 2, 3, 4, \ldots\}$. Consider the set of dots shown in Figure 1.1. The grid has infinitely many infinite rows of dots, but nevertheless we can pair each dot with a different positive integer as shown in the figure. Simply view the dots along a series of finite diagonal lines, and "count" along the successive diagonals, starting in the bottom left corner.

There is a very similar proof that the set of (positive) fractions is countable, since each fraction m/n corresponds to the pair (m, n) of positive integers. It follows that the set of positive rational numbers is countable, since each positive rational number is given by a fraction. Admittedly, there are many fractions for the same number—for example the number $1/2$ is also given by the fractions $2/4$, $3/6$, $4/8$, and so on—but we can

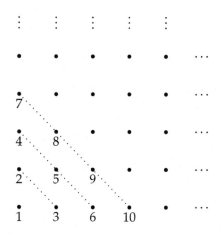

Figure 1.1. Counting the dots in an infinite grid.

list the positive rational numbers by going through the list of fractions and omitting all fractions that represent previous numbers on the list.

1.2 Does One Infinite Size Fit All?

A nice way to illustrate the elasticity of the cardinality $\aleph_0$ was introduced by the physicist George Gamow (1947) in his book *One, Two, Three, ..., Infinity*. Gamow imagines a hotel, called *Hilbert's hotel*, in which there are infinitely many rooms, numbered 1, 2, 3, 4, ... Listing the members of an infinite set is the same as accommodating the members as "guests" in Hilbert's hotel, one to each room.

The positive integers can naturally be accommodated by putting each number n in room n (Figure 1.2):

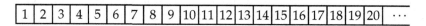

Figure 1.2. Standard occupancy of Hilbert's hotel.

The $\aleph_0$ positive integers fill every room in Hilbert's hotel, so we might say that $\aleph_0$ is the "size" of Hilbert's hotel, and that occupancy by more than $\aleph_0$ persons is unlawful. Nevertheless there is room for one more (say, the number 0). Each guest simply needs to move up one room, leaving the first room free (Figure 1.3):

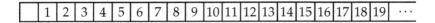

Figure 1.3. Making room for one more.

Thus $\aleph_0$ can always stretch to include one more: in symbols, $\aleph_0 + 1 = \aleph_0$. In fact, there is room for another countable infinity of "guests" (say, the negative integers $-1, -2, -3, \ldots$). The guest in room n can move to room $2n$, leaving all the odd numbered rooms free (Figure 1.4):

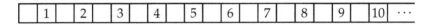

Figure 1.4. Making room for a countable infinity more.

In symbols: $\aleph_0 + \aleph_0 = \aleph_0$.

There is even room for a countable infinity of countable infinities of guests. Suppose, say, that the guests arrive on infinite buses numbered 1, 2, 3, 4, $\ldots$, and that each bus has guests numbered 1, 2, 3, 4 , $\ldots$. The guests in bus 1 can be accommodated as follows:

> put guest 1 in room 1; then skip 1 room; that is,
>
> put guest 2 in room 3; then skip 2 rooms; that is,
>
> put guest 3 in room 6; then skip 3 rooms; that is,
>
> put guest 4 in room 10; then skip 4 rooms; $\ldots$.

Thus the first bus fills the rooms shown in Figure 1.5:

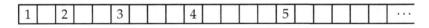

Figure 1.5. Making room for a countable infinity of countable infinities.

After the first bus has been unloaded, the unoccupied rooms are in blocks of $1, 2, 3, 4, \ldots$ rooms, so we can unload the second bus by putting its guests in the leftmost room of each block. After that, the unoccupied rooms are *again* in blocks of 1, 2, 3, 4, $\ldots$ rooms, so we can repeat the process with the third bus, and so on. (You may notice that each busload occupies a sequence of rooms numbered the same as a row in Figure 1.1.)

The result is that the whole series of $\aleph_0$ busloads, each with $\aleph_0$ guests, can be packed into Hilbert's hotel—with exactly one guest per room. In symbols: $\aleph_0 \times \aleph_0 = \aleph_0$.

The equations of "cardinal arithmetic" just obtained,

$$\aleph_0 + 1 = \aleph_0,$$
$$\aleph_0 + \aleph_0 = \aleph_0,$$
$$\aleph_0 \times \aleph_0 = \aleph_0,$$

show just how elastic the transfinite number $\aleph_0$ is. So much so, one begins to suspect that cardinal arithmetic has nothing to say except that any infinite set has cardinality $\aleph_0$. And if all transfinite numbers are the same it is surely a waste of time to talk about them. But fortunately they are *not* all the same. In particular, the set of points on the line has cardinality strictly *larger* than $\aleph_0$. Cantor discovered this difference in 1874, opening a crack in the world of the infinite from which unexpected consequences have spilled ever since. There is, after all, a lot to say about infinity, and the purpose of this book is to explain why.

1.3 Cantor's Diagonal Argument

Before studying the set of points on the line, we look at a related set that is slightly easier to handle: the set of all sets of positive integers. A set S of positive integers can be described by an infinite sequence of 0s and 1s, with 1 in the nth place just in case n is a member of S. Table 1.1 shows a few examples:

subset	1	2	3	4	5	6	7	8	9	10	11	...
even numbers	0	1	0	1	0	1	0	1	0	1	0	...
squares	1	0	0	1	0	0	0	0	1	0	0	...
primes	0	1	1	0	1	0	1	0	0	0	1	...

Table 1.1. Descriptions of positive integer sets.

Now suppose that we have $\aleph_0$ sets of positive integers. That means we can form a list of the sets, $S_1, S_2, S_3, \ldots$, whose nth member S_n is the set paired with integer n. We show that such a list can never include *all* sets of positive integers by describing a set S different from each of $S_1, S_2, S_3, \ldots$.

This is easy: for each number n, put n in S just in case n is *not* in S_n. It follows that S *differs from each S_n with respect to the number n: if n is S_n, then n is not in S; if n is not S_n, then n is in S.* Thus S is not on the list $S_1, S_2, S_3, \ldots$, and hence no such list can include all sets of positive integers.

subset	1	2	3	4	5	6	7	8	9	10	11	...
S_1	**0**	1	0	1	0	1	0	1	0	1	0	...
S_2	1	**0**	0	1	0	0	0	0	1	0	0	...
S_3	0	1	**1**	0	1	0	1	0	0	0	1	...
S_4	1	0	1	**0**	1	0	1	0	1	0	1	...
S_5	0	0	1	0	**0**	1	0	0	1	0	0	...
S_6	1	1	0	1	1	**0**	1	1	0	1	1	...
S_7	1	1	1	1	1	1	**1**	1	1	1	1	...
S_8	0	0	0	0	0	0	0	**0**	0	0	0	...
S_9	0	0	0	0	0	0	0	0	**1**	0	0	...
S_{10}	1	0	0	1	0	0	1	0	0	**1**	0	...
S_{11}	0	1	0	0	1	0	0	1	0	0	**0**	...
$\vdots$												
S	**1**	**1**	**0**	**1**	**1**	**1**	**0**	**1**	**0**	**0**	**1**	...

Table 1.2. The diagonal argument.

The argument we have just made is called a *diagonal* argument because it can be presented visually as follows. Imagine an infinite table whose rows encode the sets $S_1, S_2, S_3, \ldots$ as sequences of 0s and 1s, as in the examples above. We might have, say, the sets shown in Table 1.2.

The digit (1 or 0) that signals whether or not n belongs to S_n is set in bold type, giving a diagonal sequence of bold digits

$$00100010110\ldots.$$

The sequence for S is obtained by switching each digit in the diagonal sequence. Hence the sequence for S is necessarily different from the sequences for all of $S_1, S_2, S_3, \ldots$.

The cardinality of the set of all sequences of 0s and 1s is called $2^{\aleph_0}$. We use this symbol because there are two possibilities for the first digit in the sequence, two possibilities for the second digit, two possibilities for the third, and so on, for all the $\aleph_0$ digits in the sequence. Thus it is reasonable to say that there are $2 \times 2 \times 2 \times \cdots$ ($\aleph_0$ factors) possible sequences of 0s and 1s, and hence there are $2^{\aleph_0}$ sets of positive natural numbers.

The diagonal argument shows that $2^{\aleph_0}$ is strictly greater than $\aleph_0$ because there is a one-to-one correspondence between the positive integers and certain sets of positive integers, but not with *all* such sets. As we have just seen, if the numbers 1, 2, 3, 4, $\ldots$ are assigned to sets $S_1, S_2, S_3, S_4, \ldots$ there will always be a set (such as S) that fails to be assigned a number.

The Logic of the Diagonal Argument

Many mathematicians aggressively maintain that there can be no
doubt of the validity of this proof, whereas others do not admit it. I
personally cannot see an iota of appeal in this proof ... my mind will
not do the things that it is obviously expected to do if this is indeed
a proof.

—P .W. Bridgman (1955), p. 101

P. W. Bridgman was an experimental physicist at Harvard, and winner of
the Nobel prize for physics in 1946. He was also, in all probability, one of
the smartest people *not* to understand the diagonal argument. If you had
any trouble with the argument above, you can rest assured that a Nobel
prize winner was equally troubled. On the other hand, I do not think
that any mathematically experienced reader *should* have trouble with the
diagonal argument. Here is why.

The logic of the diagonal argument is really very similar to that of
Euclid's proof that there are infinitely many primes. Euclid faced the
difficulty that the totality of primes is hard to comprehend, since they
follow no apparent pattern. So, he avoided even considering the totality
of primes by arguing instead that *any finite list of primes is incomplete.*

Given a finite list of primes $p_1, p_2, \ldots, p_n$, one forms the number

$$N = p_1 p_2 \cdots p_n + 1,$$

which is obviously not divisible by any one of $p_1, p_2, \ldots, p_n$ (they each
leave remainder 1). But N is divisible by *some* prime number, so the list
$p_1, p_2, \ldots, p_n$ of primes is incomplete. Moreover, we can find a specific
prime p not on the list by finding the smallest number ≥ 2 that divides N.

An uncountable set is likewise very hard to comprehend, so we avoid
doing so and instead suppose that we are given a countable list $S_1, S_2,$
$S_3, \ldots$ of members of the set. The word "given" may be interpreted as
strictly as you like. For example, if $S_1, S_2, S_3, \ldots$ are sequences of 0s and
1s, you may demand a *rule* that gives the mth digit of S_n at stage $m + n$.
The diagonal argument still works, and it gives a *completely specific S* not
on the given list. (Indeed, it also leads to some interesting conclusions
about rules for computing sequences, as we will see in Chapter 3.)

The Set of Points on the Line

The goal of set theory is to answer the question of highest importance: whether
one can view the line in an atomistic manner, as a set of points.

—Nikolai Luzin (1930), p. 2.

By the "line" we mean the number line, whose "points" are known as the
real numbers. Each real number has a *decimal expansion* with an infinite

sequence of decimal digits after the decimal point. For example,

$$\pi = 3.14159265358979323846\ldots.$$

If we stick to real numbers between 0 and 1, then each number is given by the sequence of digits after the decimal point, each term of which is one of 0, 1, 2, 3, 4, 5, 6, 7, 8, or 9. A countable infinity of real numbers $x_1, x_2, x_3, x_4, \ldots$ between 0 and 1 can therefore be encoded by a table very like the table we used to encode a countable infinity of sets of positive integers. Likewise, we can construct a real number x *different* from each of $x_1, x_2, x_3, x_4, \ldots$ by arranging that the nth digit of x is different from the nth digit of x_n, for each positive integer n. As before, this amounts to looking at the diagonal digits in the table, and changing each one of them.

However, there is now a slight problem with the diagonal construction. Changing each diagonal digit certainly produces a sequence of digits different from all the sequences given for $x_1, x_2, x_3, x_4, \ldots$. But this does not ensure that the new sequence represents a new number. It could happen, for example, that the sequence obtained by the diagonal construction is

$$0 \cdot 49999999999999 \cdots$$

and that one of the given sequences is

$$x_1 = 0 \cdot 50000000000000 \cdots$$

These are different sequences, but they represent the same number, namely 1/2. Since two sequences can represent the same number only if one of them ends in an infinite sequence of 9s, we can avoid this problem by never changing a diagonal digit to a 0 or a 9. For example, we could use the following rule.

If the nth digit of x_n is 1, let the nth digit of x be 2.
If the nth digit of x_n is not 1, let the nth digit of x be 1.

With this rule, x does not merely have a sequence of digits different from those for $x_1, x_2, x_3, \ldots$. As a number, x is different from $x_1, x_2, x_3, \ldots$. Thus we have proved that *the set of real numbers is of greater cardinality than the set of positive natural numbers*. If we make a list of real numbers $x_1, x_2, x_3, x_4, \ldots$ there will always be a real number (such as x) not on the list.

In fact, the set of real numbers (whether between 0 and 1 or over the whole number line) has cardinality $2^{\aleph_0}$—the same as that of the set of sequences of 0s and 1s. The cardinality $2^{\aleph_0}$, like $\aleph_0$, measures the "size" among familiar sets in mathematics. The reasons for this will become clearer as we explore further examples of countable and uncountable sets.

1.4 Transcendental Numbers

> [On the question whether the real numbers can be made to correspond with
> the natural numbers]... if it could be answered with no, then one would have
> a new proof of Liouville's theorem that there are transcendental numbers.
> —Georg Cantor,
> Letter to Dedekind, 2 December 1873,
> in Ewald (1996), p. 844.

Cantor's discovery of uncountable sets in 1874 was one of the most un-expected events in the history of mathematics. Before 1874, infinity was not even considered a legitimate mathematical concept by most people, so the need to distinguish between countable and uncountable infinities could not have been imagined. The idea of uncountability was just *too* original to be appreciated by most mathematicians. Because of this, Cantor downplayed the idea of uncountability in the published version of his discovery, introducing it only indirectly via a "property of the algebraic numbers."

The algebraic numbers were very familiar to 19th-century mathematicians. A number x is called *algebraic* if it satisfies a polynomial equation with integer coefficients, that is, an equation of the form

$$a_n x^n + a_{n-1} x^{n-1} + \cdots + a_1 x + a_0 = 0,$$

where $a_0, a_1, \ldots, a_{n-1}, a_n$ are integers.

The algebraic numbers include all the rational numbers m/n, where m and n are integers, and also many irrational numbers, such as $\sqrt{2}$, $\sqrt{3}$, $\sqrt[3]{2}$, and so on. In fact, it is quite hard to find any numbers that are *not* algebraic. The first example was found by the French mathematician Joseph Liouville in 1844, using a result that says (roughly speaking) that an irrational algebraic number cannot be closely approximated by rationals.

Liouville then considered the number

$$x = 0 \cdot 1010010000001000000000000000000000000010\ldots,$$

where the successive blocks of 0s have lengths 1, 2, 3 × 2, 4 × 3 × 2, ...
He showed that x is irrational but that it *can* be closely approximated by rationals, hence x is not algebraic. Such numbers are now called *transcendental*, because they transcend description by algebraic means. Invariably, they have to be described by infinite processes, such as infinite decimals or infinite series. The first "well-known" number to be proved transcendental was

$$e = 1 + \frac{1}{1} + \frac{1}{1 \times 2} + \frac{1}{1 \times 2 \times 3} + \frac{1}{1 \times 2 \times 3 \times 4} + \cdots,$$

a number which makes many appearances in mathematics, perhaps most famously in the equation

$$e^{\pi\sqrt{-1}} = -1.$$

Liouville's compatriot Charles Hermite proved that e is transcendental in 1873, using some difficult calculus. In fact, Hermite was so exhausted by the effort that he gave up the idea of trying to prove that π is transcendental. The transcendance of π was first proved by the German mathematician Ferdinand Lindemann in 1882, using Hermite's methods and the above equation connecting e and π.

At any rate, in 1874 the only known approaches to transcendental numbers were those of Liouville and Hermite, using sophisticated algebra and calculus. Cantor surprised the world by showing the existence of transcendental numbers without any advanced math at all—simply by proving that *the set of algebraic numbers is countable.* Combining this with his result that real numbers are uncountable, it follows that some real numbers are transcendental. In fact, "most" real numbers must be transcendental. Not only does set theory find transcendental numbers easily, it also shows that the handful of transcendental numbers previously known actually belong to the vast, uncountable, majority.

Since we have already seen one of Cantor's proofs that there are uncountably many reals, it remains to explain why there are only countably many algebraic numbers. To do this we come back to the equations that define algebraic numbers:

$$a_n x^n + a_{n-1} x^{n-1} + \cdots + a_1 x + a_0 = 0, \tag{1}$$

where $a_0, a_1, \ldots, a_{n-1}, a_n$ are integers. The equation (1) has at most n solutions, as one learns in elementary algebra, so we can list all algebraic numbers if we can list all equations of the form (1). To do this, Cantor used a quantity called the *height* of the equation,

$$h = |a_n| + |a_{n-1}| + \cdots + |a_0| + n,$$

suggested by his colleague Richard Dedekind. It is not hard to see that *there are only finitely many equations with a given height h,* since they necessarily have degree $n \leq h$ and each coefficient has absolute value less than h. Therefore, we can list all equations (and hence all algebraic numbers) by first listing the equations of height 1, then those of height 2, those of height 3, and so on.

This listing process shows that the algebraic numbers form a countable set, and hence we are done.

1.5 Other Uncountability Proofs

We have not described Cantor's first uncountability proof, from 1874, because it is more complicated than his diagonal proof, which dates from 1891. The logic of the 1874 proof is basically the same—a countable set of numbers does not include all numbers because we can always find a number outside it—but the construction of the outsider x is not obviously "diagonal." Rather, x is a *least upper bound* of a certain increasing sequence of numbers $x_1, x_2, x_3, x_4, \ldots$; that is, x is the least number greater than all of $x_1, x_2, x_3, x_4, \ldots$. However, the least upper bound begins to look "diagonal" when one studies the decimal digits of $x_1, x_2, x_3, x_4, \ldots$.

Suppose, for the sake of example, that the numbers $x_1, x_2, x_3, x_4, \ldots$ have the following decimal digits:

$$x_1 = 1.413\ldots$$
$$x_2 = 1.4141\ldots$$
$$x_3 = 1.414232\ldots$$
$$x_4 = 1.414235621\ldots$$
$$x_5 = 1.4142356235\ldots$$
$$x_6 = 1.4142356237\ldots$$

$$\vdots$$

Because $x_1 < x_2 < x_3 < \cdots$, each decimal x_{i+1} agrees with its predecessor x_i up to a certain digit, and x_{i+1} has a *larger* digit than x_i at the first place where they disagree. These digits of first disagreement (shown in bold type) form a "jagged diagonal." And the least upper bound x of the sequence $x_1, x_2, x_3, x_4, \ldots$ is obtained by uniting all the decimal segments up to this "diagonal":

$$x = 1.4142356237\ldots.$$

Thus there is a sense in which Cantor's original uncountability proof involves a diagonal construction.

The same is true of a remarkable proof discovered by the German mathematician Axel Harnack in 1885, using the concept of *measure*. Suppose that $a_1, a_2, a_3, a_4, \ldots$ is any list of real numbers. Harnack observes that we can cover all of these numbers by line segments of total length as small as we please, say ε. Just take a line segment of length ε, break it in half, and use a segment of length $\varepsilon/2$ to cover the number a_1. Then break the remaining segment of length $\varepsilon/2$ in half and use a segment of length $\varepsilon/4$ to cover a_2, and so on. Thus we cover

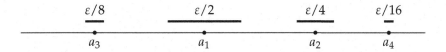

Figure 1.6. Harnack's covering of a countable set.

a_1 by an interval of length $\varepsilon/2$,
a_2 by an interval of length $\varepsilon/4$,
a_3 by an interval of length $\varepsilon/8$,
a_4 by an interval of length $\varepsilon/16$,
. . . ,

as shown in Figure 1.6. And *the whole infinite list $a_1, a_2, a_3, a_4, \ldots$ is covered by line segments of total length at most ε.*

It follows that the numbers $a_1, a_2, a_3, a_4, \ldots$ do not include all real numbers. In fact, far from filling the whole line, the set of numbers $a_1, a_2, a_3, a_4, \ldots$ has total length zero! Thus no list exhausts the set of real numbers, or even comes close. This proof shows more dramatically why there are uncountably many real numbers, but it does not seem to yield any *particular* number not in the given list $a_1, a_2, a_3, a_4, \ldots$. This defect is easily fixed, by none other than the diagonal construction. (It is also convenient to modify the lengths of the covering intervals in order to suit decimal notation.)

Suppose that the numbers $a_1, a_2, a_3, a_4, \ldots$ are given as infinite decimals, say

$$a_1 = 1 \cdot 73205\ldots,$$
$$a_2 = 0 \cdot 11111\ldots,$$
$$a_3 = 3 \cdot 14159\ldots,$$
$$a_4 = 0 \cdot 99999\ldots,$$

$$\vdots$$

If we cover a_1 by an interval of length $1/10$ then numbers x that differ from a_1 in the first decimal place by at least 2 are definitely *outside* the first covering interval. We can change the first digit in a_1 to 5, for example, so $x = 0 \cdot 5\ldots$ is outside the first interval. Next, if we cover a_2 by an interval of length $1/100$, then numbers x that differ from a_2 in the second decimal place by at least 2 are definitely outside the second covering interval. For example, we could change the second digit in a_2 to 3, so $x = 0 \cdot 53\ldots$ is outside the first and second intervals. Similarly, if we cover a_3 by an interval of length $1/1000$, then $x = 0 \cdot 533\ldots$ is outside the first, second, and third intervals.

It is clear that we can continue in this way to cover each real number a_n by an interval of length $1/10^n$, and at the same time find a number x outside all the intervals by choosing the nth digit of x to be suitably different from the nth digit of a_n. This is clearly a diagonal construction.

Thus it may be that Cantor distilled the diagonal argument from previous uncountability proofs. Indeed, a more explicitly "diagonal" construction had already been described in 1875 by another German mathematician with an interest in the infinite, Paul du Bois-Reymond. We discuss his work in the next section.

1.6 RATES OF GROWTH

From ancient times, when Archimedes tried to estimate the number of grains of sand in the universe, until today, when "exponential growth" has become a cliché, people have been fascinated by large numbers and rapid rates of growth. With modern mathematical notation it is quite easy to describe functions or sequences with extravagant growth and, with somewhat greater difficulty, to compare growth rates of different functions.

For example, the function $f(n) = n$, whose sequence of values is

$$1, \quad 2, \quad 3, \quad 4, \quad 5, \quad 6, \quad \ldots,$$

is a function that grows beyond all bounds. But it does not grow as fast as the function $g(n) = n^2$, whose values form the sequence of squares:

$$1, \quad 4, \quad 9, \quad 16, \quad 25, \quad 36, \quad \ldots.$$

We can see why by looking at $g(n)/f(n) = n$, which itself grows beyond all bounds, or *tends to infinity*, as we usually say. In general, let us agree to say that a function $G(n)$ *grows faster* than a function $F(n)$ if $G(n)/F(n)$ tends to infinity. We write this relationship symbolically as

$$G(n)/F(n) \to \infty \quad \text{as} \quad n \to \infty.$$

It follows immediately that n^3 grows faster than n^2, n^4 grows faster than n^3, and so on. Thus the infinite family of functions $\{n, n^2, n^3, n^4, \ldots\}$ represents a family of growth rates with *no greatest member*. It is called the family of *polynomial* growth rates.

Although there is no greatest polynomial rate of growth, there is a growth rate greater than all polynomial rates. Consider, for example, the growth rate of the function 2^n. This function has exponential growth, and it grows faster than n^k for any fixed k. We shall not pause to prove this fact, because there is another function that beats all of $n, n^2, n^3, n^4, \ldots$ more obviously; namely, the function

$$d(n) = n^n.$$

It is clear that

$$n^n > n^2 \quad \text{for all } n > 2,$$
$$n^n > n^3 \quad \text{for all } n > 3,$$
$$n^n > n^4 \quad \text{for all } n > 4,$$

and so on.

Thus, for any k, $n^n > n^{k+1}$ for all sufficiently large n. It follows that $n^n/n^k \to \infty$ as $n \to \infty$, because we already know that $n^{k+1}/n^k \to \infty$ as $n \to \infty$.

The values of $d(n) = n^n$ are nothing but the *diagonal values* in the table of values of the functions $n, n^2, n^3, n^4, \ldots$:

$$d(1) = 1^1 \text{ is the first value of } n,$$
$$d(2) = 2^2 \text{ is the second value of } n^2,$$
$$d(3) = 3^3 \text{ is the third value of } n^3,$$

and so on.

A similar idea applies to any sequence of functions. This is essentially what du Bois-Reymond discovered in 1875, so we name the result after him.

THEOREM OF DU BOIS-REYMOND. *If $f_1, f_2, f_3, \ldots$ is a list of positive integer functions, then there is a positive integer function that grows faster than any f_i.*

Proof: Since all function values are positive, we have

$$f_1(n) + f_2(n) + \cdots + f_n(n) > f_i(n) \quad \text{for each } i \le n.$$

The function f defined by $f(n) = f_1(n) + f_2(n) + \cdots + f_n(n)$, therefore satisfies $f(n) \ge f_i(n)$ for all $n > i$. That is, f grows *at least as fast* as any function f_i.

Consequently, if we define a function d by

$$d(n) = nf(n),$$

then $d(n)/f(n) \to \infty$ as $n \to \infty$, so d grows faster than any function f_i. $\square$

It follows that no list of functions grows fast enough to "overtake" every function of natural numbers. This was what interested du Bois-Reymond, and it is indeed a very pregnant discovery, to which we will return later. But notice also that we again have a diagonal argument for uncountability. (We regard the function $d(n) = nf(n)$ as "diagonal" because $f(n)$ is the sum of entries in or above the diagonal in the table of function values for $f_1, f_2, f_3, \ldots$.)

The set of positive integer functions is uncountable, because *for any list of such functions $f_1(n), f_2(n), f_3(n), \ldots$ there is a function $d(n)$ not in the list.*

1.7 THE CARDINALITY OF THE CONTINUUM

So far we have found three uncountable sets: the set of real numbers, the set of subsets of the positive integers, and the set of functions of positive integers with positive integer values. In a certain sense these three are essentially the *same* set, so it is not so surprising that in each case their uncountability follows by a similar argument.

Like the several countable sets discussed in Section 1.1, these three uncountable sets have the same cardinality. We called it $2^{\aleph_0}$ in Section 1.3, and it is also called the *cardinality of the continuum*, since the continuum of real numbers is the most concrete set with cardinality $2^{\aleph_0}$. We can visualize the totality of all real numbers as a continuous line—the "number line"—but it is quite hard to visualize the totality of sets of positive integers, say, until these sets have been matched up with real numbers.

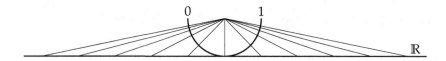

Figure 1.7. One-to-one correspondence between $\mathbb{R}$ and the interval $(0,1)$.

Before establishing a one-to-one correspondence between real numbers and sets of positive integers, we first observe that *there is a one-to-one correspondence between the set $\mathbb{R}$ of all real numbers and the interval $(0,1)$ of real numbers between 0 and 1*. This is geometrically obvious if one bends the line segment between 0 and 1 into a semicircle and then projects the semicircle onto the number line $\mathbb{R}$ as shown in Figure 1.7.

CORRESPONDENCE BETWEEN REAL NUMBERS AND SETS

A number in the interval $(0,1)$ and a set of natural numbers have a certain notational similarity. Namely, they both have natural descriptions as sequences of 0s and 1s. We have already seen how to encode a set of natural numbers by such a sequence in Section 1.2.

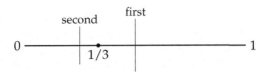

Figure 1.8. First two bisections for 1/3.

To encode a number x in $(0,1)$ by a sequence of 0s and 1s we make successive bisections of the interval, each time choosing the subinterval in which x lies. If x lies in the left half, write down 0; if the right half, write down 1. Then bisect the subinterval in which x lies, and repeat the process. The resulting sequence of 0s and 1s is called the *binary expansion* of x. For example, Figure 1.8 shows how we find the binary expansion of 1/3.

After the first bisection, 1/3 lies in the left half, so we write down 0. After bisecting the left half, 1/3 lies in the right half, so we write down 1. In the resulting quarter interval, 1/3 is in the same position as at the beginning; namely, 1/3 of the way from the left hand end of the interval. Therefore, if we continue the bisection process, we will write down 0 then 1 then 0 then 1, and so on, forever. Thus 1/3 has the infinite binary expansion

$$\cdot 010101010101010101010101\ldots$$

(where the dot is a "binary point" instead of the decimal point). Ambiguity enters when x falls on a line of bisection; for example, when $x = 1/2$ or $x = 1/4$. In that case, we can assign x to either the left or right subinterval when the line of subdivision hits x, but thereafter we have no choice. If we write 0—thus assigning x to the left half—then x will lie in the right half of every subinterval constructed thereafter, and the 0 will be followed by an infinite sequence of 1s. For example, 1/2 has the binary expansion

$$\cdot 0111111111111111111111111\ldots$$

But if we choose 1 at the beginning, then every digit thereafter is 0, so the other binary expansion of 1/2 is

$$\cdot 100000000000000000000000\ldots$$

(This is analogous to the ambiguity in decimal expansions, where 1/2 can be written as both $\cdot 499999999\ldots$ and $\cdot 500000000\ldots.$)

In general, each *binary fraction* $p/2^q$ in $(0,1)$ corresponds to two different sequences of 0s and 1s. Both sequences have the same initial segment, in one case followed by $10000\ldots$ and in the other case followed by 01111.... Thus each binary fraction corresponds to two different sets of natural numbers. Fortunately, there are only countably many binary fractions (for example, because they form a subset of the set of rationals) so this breakdown of the one-to-one correspondence is easily fixed.

In fact we have already seen, in Section 1.1, how countable sets in two-to-one correspondence are also in one-to-one correspondence (consider the set of all natural numbers and the set of all even numbers). Thus the two-to-one correspondence between sequences ending in 10000... or 01111... and binary fractions can be rearranged into a one-to-one correspondence. Combining this with the one-to-one correspondence between the remaining sequences of 0s and 1s and the remaining numbers between 0 and 1 gives the required one-to-one correspondence between sets of natural numbers and real numbers between 0 and 1.

Correspondence between Functions and Real Numbers

Each function f on the set of positive integers, with positive integer values, corresponds to a real number between 0 and 1 as follows. First write down the sequence of values of f, say,

$$4, \quad 2, \quad 6, \quad 1, \quad 1, \quad 8, \quad 3, \quad 5, \quad \dots \quad .$$

Next, encode this sequence by a sequence of 0s and 1s, replacing each positive integer $f(n)$ by a block of $f(n) - 1$ (hence possibly zero) 1s, and each comma by a 0:

$$1110101111100011111110110111110\dots.$$

Finally, insert a binary point in front to make this sequence the binary expansion of a real number. The expansion is necessarily one with infinitely many 0s, because the function f has infinitely many values. But this is fine, because it means we omit all binary expansions ending in 1111..., and hence we get each real number between 0 and 1 at most once.

Conversely, each real between 0 and 1 has a unique binary expansion with infinitely many 0s, and there is a unique function f encoded by this expansion. For example, if the binary expansion is

$$\cdot 0010110011111011010101\dots.$$

then the successive values of f are

$$1, \quad 1, \quad 2, \quad 3, \quad 1, \quad 6, \quad 3, \quad 2, \quad 2, \quad 2, \quad 2, \quad \dots \quad .$$

In general,

$f(1) = $ (number of 1s before the first 0) $+ 1$,

$f(2) = $ (number of 1s after the first 0 and before the second 0) $+ 1$,

$f(3) = $ (number of 1s after the second 0 and before the third 0) $+ 1$,

and so on.

Thus binary expansions give a one-to-one correspondence between positive integer functions and real numbers between 0 and 1.

1.8 Historical Background

Infinity in Ancient Greece

> Zeno's argument makes a false assumption in asserting that it is impossible for a thing to pass over or severally to meet with infinite[ly many] things in finite time.
>
> —Aristotle
> *Physics*, Book VI, Chap. 2.

Since ancient times, infinity has been a key part of mathematics, though its use has often been considered harmful. Around 500 BCE, the Pythagoreans discovered the irrationality of $\sqrt{2}$, thus beginning a long struggle to grasp (what we now call) the concept of real number. This was part of a larger struggle to reconcile the continuous magnitudes arising in geometry—length, area, volume—with the discrete natural numbers $1, 2, 3, 4, \ldots$ arising from counting.

The shock of irrationality apparently left the Greeks unwilling to treat continuous magnitudes as numbers. For example, they viewed the product of two lengths as a rectangle, the product of three lengths as a box, and the product of four lengths as having no meaning at all. Nevertheless, they made great progress in *relating* continuous magnitudes to the natural numbers. Around 350 BCE, Eudoxus introduced a "theory of proportion" (known to us from Book V of Euclid's *Elements*), which was as close as anyone came to defining real numbers before the 19th century. Euclid characterized irrational lengths l as those for which a certain process (the euclidean algorithm on l and 1) is infinite. Archimedes found areas and volumes of curved figures by infinite processes, even breaking the taboo against "actual" infinite sets in some of his methods.

Almost all mathematicians until the 19th century made a sharp distinction between what they called the *potential* and *actual* infinite. The potential infinite is typically an *unending process,* such as counting $1, 2, 3, \ldots$ or cutting a curved region into smaller and smaller triangles. The actual

20 1. The Diagonal Argument

infinite is the supposed *completion* of an unending process, such as the set $\{1,2,3,\ldots\}$ of all natural numbers. In principle, it is contradictory to speak of the completion of an unending process, but in practice many processes beg for completion, because they have a clear *limit*.

The famous paradoxes of Zeno, such as Achilles and the tortoise, involve processes with a clear limit. Achilles starts behind the tortoise, but runs faster, so clearly he will catch up with the tortoise at some point. The worry, at least for Zeno, is that Achilles is behind the tortoise *at infinitely many stages* (the first stage ends when Achilles reaches the tortoise's starting point, the second when Achilles completes the tortoise's first stage, the third when Achilles completes the tortoise's second stage, and so on). So, apparently, motion involves completing an infinite sequence of events. For the Greeks, this made the concept of motion seem problematic, and they looked for ways to resolve the paradox in terms of potential infinity (as Aristotle did in his *Physics*). For us, it probably supports the idea that actual infinity exists.

At any rate, in ancient Greek mathematics there are many examples where the limit of an infinite process gives interesting new knowledge. For example, Euclid and Archimedes found that both the volume of the tetrahedron and the area of a parabolic segment may be found from the infinite series

$$1 + \frac{1}{4} + \frac{1}{4^2} + \frac{1}{4^3} + \frac{1}{4^4} + \cdots,$$

which has sum $4/3$. They are able to find this sum by considering only the potential infinity of terms

$$1, \quad 1 + \frac{1}{4}, \quad 1 + \frac{1}{4} + \frac{1}{4^2}, \quad 1 + \frac{1}{4} + \frac{1}{4^2} + \frac{1}{4^3}, \quad 1 + \frac{1}{4} + \frac{1}{4^2} + \frac{1}{4^3} + \frac{1}{4^4}, \quad \cdots$$

and showing that

- each of these terms is less than $4/3$, and

- each number less than $4/3$ is exceeded by some term in the sequence.

Thus it is fair to call $4/3$ the sum of $1 + \frac{1}{4} + \frac{1}{4^2} + \frac{1}{4^3} + \frac{1}{4^4} + \cdots$, because we have exhausted all other possibilities. This *method of exhaustion*, also invented by Eudoxus, can be used to avoid actual infinities in many of the cases that interested the Greeks. *But not all.*

Archimedes wrote a book called *The Method*, which was lost for many centuries and did not influence the later development of mathematics. It describes the method by which he discovered some of his most famous results, such as the theorem that the volume of a sphere is $2/3$ the volume of its circumscribing cylinder. The book resurfaced around 1900, and only then was it realized that Archimedes had used actual infinity in a way that cannot be avoided. In finding certain volumes, Archimedes views a solid body as a sum of *slices of zero thickness*. As we now know, there are uncountably many such slices—corresponding to the uncountably many points on the line—so one cannot view the sum of all slices as the "limit" of finite sums. Of course, it is unlikely that Archimedes had any inkling of uncountability, though he may have suspected that he was dealing with a new kind of infinity.[1]

The Method gives evidence that *intuition about infinity*, even about the continuum, exists and is useful in making mathematical discoveries. More fruits of this intuition appeared around 1800.

The First Modern Intuitions About the Continuum

... we ascribe to the straight line completeness, absence of gaps, or continuity. In what then does this continuity consist?

—Richard Dedekind
Continuity and Irrational Numbers, Chapter III,
in Dedekind (1901), p. 10.

The fundamental property of the continuum $[0, 1]$ is that it is, well, *continuous*, in the sense that it *fills the space between its endpoints without gaps*. Around 1800, the German mathematician Carl Friedrich Gauss realized that this seemingly obvious property is the key to a difficult theorem that several mathematicians had vainly sought to prove—the so-called "fundamental theorem of algebra." This theorem states that any polynomial equation, such as

$$x^4 - 2x^2 + 3x + 7 = 0, \quad \text{or} \quad x^5 - x + 1 = 0,$$

is satisfied by some complex number x. Gauss himself had trouble proving the theorem, and all the proofs he offered are incomplete by modern standards. However, in 1816 he gave a proof that clearly identifies the difficulty: it all boils down to the absence of gaps in the continuum.

Gauss's 1816 proof takes any polynomial equation and reduces it, by purely algebraic manipulations, to an equation of *odd degree*, say $p(x) = 0$. This means that the highest-degree term in $p(x)$, such as x^5 in the second example above, is an odd power of x. Now, for large values of x, the polynomial $p(x)$ is dominated by its highest-degree term, and so $p(x)$ *changes sign* between large negative values of x and large positive values of x, since this happens to any odd power of x.

Thus the graph $y = p(x)$ is a continuous curve which moves from negative values of y to positive values. Figure 1.9 shows the curve for

---

[1] For an up-to-date report on *The Method*, with some interesting mathematical speculations, see Netz and Noel (2007).

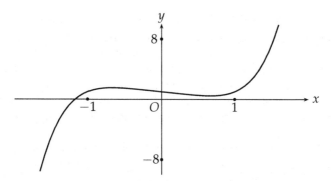

Figure 1.9. Graph of the polynomial $y = x^5 - x + 1$.

the actual example $y = x^5 - x + 1$. Since the curve passes from below to above the x-axis, and the x-axis has no gaps, the curve necessarily *meets* the x-axis.

That is, *there is a real value of x for which $x^5 - x + 1 = 0$.* Similarly, any odd-degree polynomial equation has a real solution, and the fundamental theorem of algebra follows. (Gauss's reduction to odd degree also involves solving quadratic equations, which we know have solutions, sometimes in the complex numbers.)

Gauss's intuition had led him to assume what we now call the *intermediate value theorem*: any continuous function $f(x)$ that takes both positive and negative values between $x = a$ and $x = b$ takes the value zero for some $x = c$ between a and b. The first to identify this assumption, and to attempt to prove it, was the Czech mathematician Bernard Bolzano in 1816. Bolzano was ahead of his time, not only in noticing a property of continuous functions in a theorem previously thought to belong to algebra, but also in realizing that the intermediate value property depends on the nature of the continuum.

Bolzano's attempted proof was incomplete, because a definition of the continuum was completely lacking in his time. However, he correctly identified a *completeness condition* that any reasonable concept of continuum must satisfy. This is the *least upper bound property* : if S is a set of real numbers with an upper bound, then S has a least upper bound. (That is, among the numbers greater than or equal to all members of S, there is a least.)

In 1858, Richard Dedekind brought this train of thought to a satisfying conclusion, defining real numbers by what he called *cuts* in the rationals. A cut is intuitively a separation of the rational numbers into a *lower set L* and an *upper set U*, as if by an infinitely sharp knife. Formally, a cut is a pair (L, U) where L and U are sets that together are all the rationals and such that every member of L is less than every member of U. Cuts (L, U) represent both rational and irrational numbers as follows:

- If L has a greatest member, or U has a least member, say r, then (L, U) represents the rational number r.

- If L has no greatest member and U has no least member, then (L, U) represents an irrational number. (This happens, for example, when U consists of the positive rationals with square > 2, and L consists of the remaining rationals. The pair (L, U) then represents the irrational number we call $\sqrt{2}$.)

The latter type of cut represents a *gap* in the rationals and, at the same time, provides an object to fill the gap—the cut (L, U). With breathtaking nerve, Dedekind created a gapless continuum by filling each gap in the rationals, *taking the object that fills each gap to be essentially the gap itself.*[2]

INFINITE DECIMALS REVISITED

It should be mentioned that the infinite decimals, used to model real numbers earlier in this chapter, are essentially a more readable version of Dedekind cuts. An infinite decimal, such as 3.14159..., represents a cut in the less crowded set of *decimal fractions*, separating the nearest neighbors less than 3.14159...,

$$3, \quad 3.1, \quad 3.14, \quad 3.141, \quad 3.1415, \quad 3.14159, \quad \ldots,$$

from the nearest neighbors greater than 3.14159... ,

$$4, \quad 3.2, \quad 3.15, \quad 3.142, \quad 3.1416, \quad 3.14160, \quad \ldots.$$

Infinite decimals are easy to read and understand, but try defining their sum and product! You will probably fall back on adding and multiplying their neighboring decimal fractions, much as with Dedekind cuts.

Of course, it is not enough that there are no gaps in the set of cuts. We also need to know that *cuts are entities that behave like numbers.* This is true. One can define the "sum" and "product" of two cuts in terms of the rational numbers in them, and this "sum" and "product" have the usual algebraic properties. For example, the cut for $\sqrt{2} + \sqrt{3}$ has lower

---

[2]Dedekind slightly tarnished the purity and boldness of his idea by insisting on his right to *create a new object* to fill each gap. It is perfectly valid, and more economical, to insist that the gap itself is a genuine mathematical object, which we can take to be the pair (L, U).

set consisting of all the numbers $r + s$, where r is in the lower set for $\sqrt{2}$ and s is in the lower set for $\sqrt{3}$. This is the same as the cut for $\sqrt{3} + \sqrt{2}$ because $r + s = s + r$. Best of all, *any bounded set S of cuts has a least upper bound*. The least upper bound of S has a lower set obtained by uniting the lower sets for all members of S.

Thus, with Dedekind's definition of real numbers, it was finally possible to prove the intermediate value theorem, and hence the fundamental theorem of algebra. At the same time, the proof initiated a new direction in mathematical thought. Previously undefined mathematical objects became defined in terms of sets, and every set became a legitimate mathematical object—even the uncountable set of real numbers.

Indeed, the set of real numbers was welcomed by many mathematicians as a mathematical model of the line. Today, the "number line" seems like a simple idea, but it is not! A "point" is a whole universe to someone who knows that it is actually a cut in the infinite set of rational numbers. Nonetheless, in the 1870s, many mathematicians saw this *arithmetization of geometry* as the best way to build the foundations of mathematics. Arithmetization resolved the ancient conflict between numbers and geometric magnitudes; it also provided a common foundation for geometry and calculus. And arithmetization was timely because Cantor had just started exploring the set concept itself.

However, further exploration of the set concept led to some surprises.

The Paradoxes of Set Theory

...a mere practical joke played on mankind by the goddess of wisdom.
—Azriel Levy (1979), p. 7.

The function-based diagonal argument of Paul du Bois-Reymond (Section 1.6) had far-reaching consequences, as we will see in later chapters of this book. The set-based diagonal argument of Cantor (Section 1.3) had consequences that were more immediate and dramatic.

In 1891, Cantor realized that the diagonal argument applies to *any* set, showing that *any set X has more subsets than members*. Of course, we cannot generally visualize a tabulation of subsets of X, as we did in Section 1.3 for subsets of the natural numbers. It is sufficient to consider any *one-to-one correspondence* between members x of X and subsets of X. Let S_x be the subset corresponding to x. From the sets S_x we define the "diagonal set" S, whose members are the x such that x does *not* belong to S_x.

It is clear that S *differs from each set S_x* with respect to the element x: if x is in S_x then x is not in S, and if x is not S_x then x is in S. Thus, any pairing of subsets of X with members of X fails to include all subsets. This is what we mean by saying that X has more subsets than members.

It follows, in turn, that *there is no largest set*. And, therefore, *there is no such thing as the "set of all sets"*—because the set of all sets, if it exists, is necessarily the largest set. When Cantor noticed this, in 1895, it seems to have given him some pause. What exactly does the word "set" mean if there is no set of all sets? Cantor did not have a precise answer to this question, but neither was he greatly perturbed. He had no commitment to the "set of all sets" and was content to consider only sets obtained by clear operations on given sets, such as collecting all subsets.

The question was more troubling for philosophers of mathematics, such as Gottlob Frege and Bertrand Russell, who believed that any property $\mathcal{P}$ should determine a set—the set of all objects with property $\mathcal{P}$. When the property in question is "being a set" then this belief leads to the "set of all sets." Indeed, Russell rediscovered the contradiction in the "set of all sets" in 1901, in a form that became famous as the *Russell paradox*. Russell's contribution was to distill the contradiction into "the set of all sets that are not members of themselves." The latter "set" R is immediately self-contradictory, since R belongs to R if and only if R does *not* belong to R.

Russell's argument convinced mathematicians that the set concept needs clarification, and it reinforced the idea from the 1870s that mathematics needs secure foundations. This "crisis in foundations" (and other "crises" we will meet later) had profound consequences, as we will see in the rest of the book. The problem facing set theory was described by the German mathematician Ernst Zermelo (1908) as follows.

> ...the very existence of this discipline seems to be threatened by certain contradictions ... In particular, in view of the "Russell antinomy" of the set of all sets that do not contain themselves as elements, it no longer seems admissible today to assign to an arbitrary logically defined notion a set, or class, as its extension.

Zermelo believed that set theory could be saved by *axioms for sets*, formalizing Cantor's intuition that all sets arise from given sets (such as the set of natural numbers) by well-defined operations (such as collecting all subsets).

The axioms for sets most commonly used today are due to Zermelo (1908), with an important supplement from his compatriot (who later moved to Israel) Abraham Fraenkel in 1922. Because of this they are called the *ZF axioms*. They are written in a formal language for set theory that I have not introduced, but most of them can be clearly expressed in ordinary language with the help of the concept of "membership." We write the set with members $a, b, c, \ldots$ as $\{a, b, c, \ldots\}$. Thus the brackets "comprehend" the objects $a, b, c, \ldots$ (which may themselves be sets) as members of a set.

Axiom 1. Two sets are equal if and only if they have the same members.

Axiom 2. There is a set with no members, called the *empty set*.

Axiom 3. For any sets X and Y, there is a set whose only members are X and Y. (This set, $\{X, Y\}$, is called the *unordered pair* of X and Y. Note that, when $Y = Z$, Axiom 1 gives $\{Y, Z\} = \{Y\}$. Thus the pairing axiom also gives us the "singleton" set $\{Y\}$ whose single member is Y.) The ZF axioms are the following.

Axiom 4. For any set X there is a set whose members are the members of members of X. (In the case where $X = \{Y, Z\}$, the members of members of X form what is called the *union of Y and Z*, denoted by $Y \cup Z$. In all cases, the set of members of members of X is called the *union* of the members of X.)

Axiom 5. For any set X, there is a set whose members are the *subsets* of X, where a subset of X is a set whose members are members of X. (The set of subsets of X is called the *power set* of X.)

Axiom 6. For any function definition f, and set X, the values $f(x)$, where x is a member of X, form a set. (This set is called the *range* of the function f on the *domain* X, and the axiom is called *replacement*.)

Axiom 7. Any nonempty set X has a member Y with no members in X. (A more enlightening version of this axiom—though harder to express in formal language—is the following. *There is no infinite descending sequence for set membership.* That is, if one takes a member X_1 of X, then a member X_2 of X_1, and so on, then this process can continue for only finitely many steps.)

Axiom 8. There is an infinite set, in fact a nonempty set which, along with any member X, also has the member $X \cup \{X\}$.

Axioms 1–6 say that sets are built from the empty set by operations of pairing, union, power, and replacement (taking the range of a function). To say that a function f is "defined" means that f is expressed by a formula in the formal language of ZF, which basically consists of logic symbols and symbols for membership and equality. We study formal languages for mathematics in Chapters 3, 4, and 5.

The mysterious Axiom 7, called the *axiom of foundation*, enables us to prove that *every* set arises from the empty set by the operations above. It may seem that we have a severely limited menu of sets, but in fact there are sets that can play the roles of all the objects normally needed for mathematics. First, following an idea of the Hungarian (later American) mathematician John von Neumann from the 1920s, we can define the natural numbers as follows. Let 0 be the empty set (what else?), then let

$$1 = \{0\},$$
$$2 = \{0, 1\},$$
$$3 = \{0, 1, 2\},$$

and so on. Notice that we have $n + 1 = n \cup \{n\}$ and $m < n$ if and only if m is a member of n. Thus set theory gives us definitions of the successor function and the "less than" relation for free, and we can develop arithmetic.

Next, Axiom 8 (the *axiom of infinity*), together with Axiom 6, tells us that there is a set whose members are $0, 1, 2, 3, \ldots$, so we have the set of natural numbers. Taking its power set, we are well on the way to the real numbers, the number line, geometry, calculus, and virtually everything else.

Who knew that the empty set could be so fruitful?

- Chapter 2 -

Ordinals

Preview

In the previous chapter we studied sets that can be "listed" completely, with every member in a positive integer position. We found that some sets cannot be listed in this way, even though the list is infinite. The diagonal argument always gives a new member. Can we continue the list past infinity?

Georg Cantor counted past infinity using the concept of *ordinal numbers*. The natural numbers 0, 1, 2, 3, ... are the finite ordinal numbers, but they have a least upper bound ω—the first *transfinite* ordinal number. A neat way to formalize this is to let ω be the set $\{0, 1, 2, 3, \ldots\}$. Then ω is the first term beyond the finite ordinals on the list $0, 1, 2, 3, \ldots, \omega$, ordered by membership.

Like the finite ordinals, each transfinite ordinal has a successor, so there are ordinals $\omega + 1, \omega + 2, \omega + 3$, and so on.

Ordinals keep pace with the production of new objects by the diagonal argument, because any countable set of ordinals has a least upper bound. Beyond $\omega + 1, \omega + 2, \omega + 3, \ldots$ there is $\omega \cdot 2$. Beyond $\omega \cdot 2, \omega \cdot 3, \omega \cdot 4, \ldots$ there is ω^2. Beyond $\omega^2, \omega^3, \omega^4, \ldots$ there is ω^ω. And these are merely a few of what we call the *countable ordinals*.

The least upper bound of the countable ordinals is the *first uncountable ordinal*, called ω_1. Thus ordinals offer a different road to uncountable infinity. The ordinal road is slower, but more—shall we say—orderly.

Not only are ordinals ordered, they are *well*-ordered. Any nonempty set of them has a least member; equivalently, there is no infinite descending sequence of ordinals. Thus, *even though the road to an ordinal may be long, any return trip is short* (because it is finite). This fact has surprising consequences in the world of finite objects, as we show in Sections 2.7 and 2.8.

2.1   Counting Past Infinity

When is it reasonable to count 1, 2, 3, … infinity? One such occasion is when we try to order functions according to their rate of growth. As we saw in Section 1.6, it is easy to find a sequence $f_1, f_2, f_3, \ldots$ of integer functions, each of which grows faster than the one before. We also saw, in this situation, that the diagonal function $f(n) = f_n(n)$ grows faster than *all* of $f_1(n), f_2(n), f_3(n), \ldots$. Thus it is natural to label the diagonal function f by a subscript "number" greater than all the numbers 1, 2, 3, … . Following Cantor, we use the symbol ω for this "number." Thus $f_\omega(n) = f_n(n) = n^n$ when $f_k(n) = n^k$.

The symbol ω is *omega*, the last letter of the Greek alphabet, but it belies its name because it is not the last number. In fact ω is the first *transfinite ordinal*, the least ordinal greater than all of 1, 2, 3, …, which are also known as the *finite ordinals*. The ordinal numbers

$$1, \quad 2, \quad 3, \quad \ldots, \quad \omega$$

have an immediate successor, called $\omega + 1$. It can be represented by the integer function

$$f_{\omega+1}(n) = n^{n+1},$$

which grows faster than any of $f_1, f_2, f_3, \ldots, f_\omega$. The function $f_{\omega+1}$ has greater growth rate than f_ω because $f_{\omega+1}(n)/f_\omega(n) = n \to \infty$ as $n \to \infty$. In fact we need a whole infinite sequence of increasingly large ordinals

$$\omega + 1, \quad \omega + 2, \quad \omega + 3, \quad \ldots$$

to label the increasingly fast-growing functions

$$f_{\omega+1}(n) = n^{n+1}, \quad f_{\omega+2}(n) = n^{n+2}, \quad f_{\omega+3}(n) = n^{n+3}, \quad \ldots \quad .$$

And of course, the diagonal construction gives a function f that grows faster than any of the functions $f_{\omega+1}(n) = n^{n+1}, f_{\omega+2}(n) = n^{n+2}, f_{\omega+3}(n) = n^{n+3}, \ldots$; namely,

$$f(n) = n^{2n}.$$

We therefore need to label f with an ordinal number greater than all of

$$1, \quad 2, \quad 3, \quad \ldots, \quad \omega, \quad \omega + 1, \quad \omega + 2, \quad \omega + 3, \quad \ldots \quad .$$

We call the least such ordinal $\omega \cdot 2$. (It could equally well be called $2 \cdot \omega$, but we don't, because there is a concept of product of ordinals and it so happens that $2 \cdot \omega \neq \omega \cdot 2$.)

As we know, du Bois-Reymond's theorem allows us to find a function g that grows faster than any function on a list $g_1, g_2, g_3, \ldots$. To label increasing growth rates in an orderly manner we therefore need a least upper bound for any set of ordinal numbers—a least ordinal greater than or equal to all members of the set. So far we have

$$\omega = \text{least upper bound of } \{1, 2, 3, \ldots\}$$
$$\omega \cdot 2 = \text{least upper bound of } \{\omega + 1, \omega + 2, \omega + 3, \ldots\}.$$

We also know that, if $g(n)$ is any increasing function, then $ng(n)$ is an increasing function with greater growth rate. Thus if $g(n)$ is labelled by the ordinal α we need a *successor ordinal* $\alpha + 1$ to label $ng(n)$. Starting with our biggest ordinal so far, $\omega \cdot 2$, the successor operation gives

$$\omega \cdot 2 + 1, \quad \omega \cdot 2 + 2, \quad \omega \cdot 2 + 3, \quad \ldots,$$

and the least upper bound operation gives an ordinal greater than any of these, which is naturally called $\omega \cdot 3$. This kicks off another infinite sequence

$$\omega \cdot 3 + 1, \quad \omega \cdot 3 + 2, \quad \omega \cdot 3 + 3, \quad \ldots,$$

with least upper bound $\omega \cdot 4$, followed by the sequence

$$\omega \cdot 4 + 1, \quad \omega \cdot 4 + 2, \quad \omega \cdot 4 + 3, \quad \ldots,$$

with least upper bound $\omega \cdot 5$, and so on. Doing this over and over we also get $\omega \cdot 6$, $\omega \cdot 7$, $\omega \cdot 8$, $\ldots$ —*another* increasing infinite sequence of ordinals. You will not be surprised to hear that the least upper bound of

$$\omega, \quad \omega \cdot 2, \quad \omega \cdot 3, \quad \ldots$$

is called $\omega \cdot \omega = \omega^2$. (Or that our operations for producing functions give $f_{\omega^2}(n) = n^{n^2}$.)

Where Is All This Going?

Starting with the ordinal ω^2, we can climb past

$$\omega^2 + 1, \quad \omega^2 + 2, \quad \omega^2 + 3, \quad \ldots \quad \text{to} \quad \omega^2 + \omega,$$

and similarly past

$$\omega^2 + \omega \cdot 2, \quad \omega^2 + \omega \cdot 3, \quad \omega^2 + \omega \cdot 4, \quad \ldots, \quad \text{to} \quad \omega^2 + \omega^2 = \omega^2 \cdot 2.$$

Similar long climbs lead successively to

$$\omega^2 \cdot 3, \quad \omega^2 \cdot 4, \quad \omega^2 \cdot 5, \quad \ldots,$$

whose least upper bound we naturally call ω^3. Further climbs lead successively to $\omega^4, \omega^5, \omega^6, \ldots$, whose least upper bound is naturally called ω^ω. After that (a *long* time after), we reach the dizzy heights of

$$\omega^{\omega^\omega}, \quad \omega^{\omega^{\omega^\omega}}, \quad \omega^{\omega^{\omega^{\omega^\omega}}}, \quad \ldots,$$

whose least upper bound might be denoted $\omega^{\omega^{\omega^{\omega^{\cdot^{\cdot^{\cdot}}}}}}$. At this point we are reaching the limits of standard mathematical notation, so we abbreviate the symbol $\omega^{\omega^{\omega^{\omega^{\cdot^{\cdot^{\cdot}}}}}}$ by ε_0. The name ε_0 ("epsilon zero") was chosen by Cantor, and it is slightly ironic, because the Greek letter ε is traditionally used by mathematicians to denote a "small" quantity.

Actually, ε_0 *is* small compared with what comes after it. We can clearly continue the process of generating ordinal numbers indefinitely, though perhaps with some difficulty in finding names for them and in identifying the corresponding positive integer functions. (The function $f_{\varepsilon_0}(n)$ also challenges our notational resources, because it is the diagonal function of the functions

$$f_\omega(n) = n^n, \quad f_{\omega^\omega}(n) = n^{n^n}, \quad f_{\omega^{\omega^\omega}}(n) = n^{n^{n^n}}, \quad \ldots,$$

as you might like to verify.) The more fundamental question is: what does it mean to "continue indefinitely" the process of generating ordinal numbers? We know better than to think that all infinities are the same, so we need to formulate a more precise question. At least three questions come to mind:

1. Can the current process for generating ordinal numbers (taking successors and least upper bounds) be continued *uncountably* far?

2. If so, can we "exhibit" an uncountable set of positive integer functions, ordered by increasing growth rate?

3. Do the corresponding growth rates eventually exceed that of any given positive integer function?

These questions are still a little vague, but we will give a precise answer to Question 1 in the next section. Questions 2 and 3 also have answers (sort of) but they are entangled with deep questions of set theory we are able to discuss only briefly in this book (Sections 2.3 and 2.4).

Nevertheless, simply trying to understand these questions has been amazingly fruitful in the development of logic and mathematics, as we will see in later chapters.

2.2   THE COUNTABLE ORDINALS

In Chapter 1 we discussed cardinal numbers, though we did not need to define precisely what a cardinal number is. It is enough to know when two sets A and B have the *same* cardinal number, namely, when there is a one-to-one correspondence between the elements of A and B. Likewise, it is possible to say what it means for sets A and B to have the same ordinal number, though this is somewhat more complicated.

First, it is necessary for A and B to be *well-ordered*. A is said to be *linearly ordered* by a relation $<_A$ on its members if the following conditions hold.

1. For any member a of A, a is not $<_A a$.

2. For any two members a_1 and a_2 of A, either $a_1 <_A a_2$ or $a_2 <_A a_1$.

3. For any members a_1, a_2, a_3 of A, if $a_1 <_A a_2$ and $a_2 <_A a_3$ then $a_1 <_A a_3$.

Examples of linearly ordered sets are the real numbers and the positive integers, under the usual $<$ relation for numbers. However, only the positive integers are *well*-ordered by the $<$ relation. An order relation $<_A$ is called a *well-ordering* of a set A if it satisfies the additional condition:

4. Any nonempty subset of A has a $<_A$-least element, that is, an element l such that $l <_A a$ for any a in A other than l. This implies that *there is no infinite descending sequence* $a_1 >_A a_2 >_A a_3 >_A \cdots$ of members of A, otherwise the subset $\{a_1, a_2, a_3, \ldots\}$ of A has no $<_A$-least member.

The real numbers are not well-ordered by $<$ because, for example, the set of numbers greater than 0 has no least member. It is obvious that any set of positive integers has a least member. With only a little more thought, one sees that the ordinal numbers we wrote down in the previous section are also well-ordered by the left-to-right order in which we wrote them down. For example, consider the ordinals less than $\omega \cdot 2$:

$$1, \quad 2, \quad 3, \quad \ldots \quad \omega, \quad \omega + 1, \quad \omega + 2, \quad \ldots.$$

It is clear that any nonempty subset S of these ordinals has a leftmost member. If S includes any positive integers, then the least of these is leftmost. If S includes only elements among the ordinals $\omega + m$, for ordinary integers m, then the leftmost element is the one for which the integer m is least. A similar argument applies for the sets of ordinals up to any ordinal we mentioned so far, though naturally the argument becomes more complicated as larger ordinals are included.

Here is an example of a well-ordering—different from the standard one—of the set of all positive integers. We make the following picture,

$$1, \quad 3, \quad 5, \quad \ldots \quad 2, \quad 4, \quad 6, \quad \ldots$$

and say that number m is "less than" number n if m is further to the left in the picture. That is, m is "less than" n if either m is odd and n is even, or if they are both even or both odd and $m < n$ in the usual sense. This ordering of positive integers is in fact the *same* as the ordering of ordinals less than $\omega \cdot 2$, in a sense we now make precise.

Two well-ordered sets A and B have the *same ordinal number* if there is a one-to-one correspondence f between A and B that preserves the ordering. That is, if a_1 and a_2 are two members of A then $f(a_1)$ and $f(a_2)$ are two members of B, any member b of B equals $f(a)$ for some a in A, and

$$a_1 <_A a_2 \quad \text{if and only if} \quad f(a_1) <_B f(a_2).$$

The two orderings

$$1, \quad 2, \quad 3, \quad \ldots \quad \omega, \quad \omega + 1, \quad \omega + 2, \quad \ldots$$

and

$$1, \quad 3, \quad 5, \quad \ldots \quad 2, \quad 4, \quad 6, \quad \ldots$$

clearly have the "same ordinal number" in the sense of this definition, and indeed you probably have the urge to say the common ordinal number is $\omega \cdot 2$. This is correct, though a full explanation requires a definition of the concept "ordinal number," which is more complicated than merely defining the relation "same ordinal number as." However, we indicate how the idea may be justified by the axioms for ZF set theory listed in Section 1.8.

The first ordinal is the empty set, which we write $\{\}$ to make it plain that it has no members. It is an axiom of set theory that the empty set exists. As mentioned in Section 1.8, John von Neumann had the brilliant idea to let 0 be the empty set, and to define the finite ordinals successively as the following sets:

$$1 = \{0\}, \quad 2 = \{0,1\}, \quad 3 = \{0,1,2\}, \quad \ldots.$$

In general, the finite ordinal $n + 1 = \{0, 1, 2, \ldots, n\}$ is the set of all predecessors of $n + 1$, so the ordinals less than the set $n + 1$ are precisely its *members*. Conversely, the *successor* $n + 1$ of the ordinal n is simply the set $n \cup \{n\}$, whose members are n and the members of n. Thus we can build all the finite ordinals from "nothing" (the empty set) by taking old sets as members of new sets, and *we can order them by the membership relation*.

It doesn't stop there. Another axiom of set theory (the axiom of infinity) says that we can collect all the finite ordinals in a single set

$$\omega = \{0, 1, 2, 3, \ldots\}.$$

Thus the predecessors of the first infinite ordinal, ω, are precisely its members. Then the successor operation gives

$$\omega + 1 = \omega \cup \{\omega\} = \{0, 1, 2, \ldots, \omega\},$$
$$\omega + 2 = \omega + 1 \cup \{\omega + 1\} = \{0, 1, 2, \ldots, \omega, \omega + 1\},$$
$$\omega + 3 = \omega + 2 \cup \{\omega + 2\} = \{0, 1, 2, \ldots, \omega, \omega + 1, \omega + 2\},$$
$$\ldots.$$

And we can (with the help of the replacement axiom) collect all the ordinals so far into the set

$$\omega \cdot 2 = \{0, 1, 2, \ldots, \omega, \omega + 1, \omega + 2, \ldots\}.$$

Thus the ordinal $\omega \cdot 2$ represents the left-to-right ordering of its members

$$0, \quad 1, \quad 2, \quad \ldots \quad \omega, \quad \omega + 1, \quad \omega + 2, \quad \ldots \quad,$$

which is the same as the left-to-right ordering of positive integers

$$1, \quad 3, \quad 5, \quad \ldots \quad 2, \quad 4, \quad 6, \quad \ldots \quad,$$

as we surmised above. All of the ordinals considered in the previous section can be shown to represent well-orderings of the positive integers. Another example is $\omega \cdot 3$, which represents the left-to right ordering

$$1, \quad 4, \quad 7, \quad \ldots \quad 2, \quad 5, \quad 8, \quad \ldots \quad 3, \quad 6, \quad 9, \quad \ldots \quad .$$

Conversely, any well-ordering of the positive integers is represented by what we call a *countable ordinal*—a countable set α with properties that generalize those of the particular ordinals mentioned above: α *is well-ordered by the membership relation*, and *any member of a member of α is a member of α*.

The countable ordinals go inconceivably far beyond the ordinal ε_0 that we struggled to reach in the last section. Any countable ordinal α has a countable successor $\alpha + 1$, and any countable increasing sequence of countable ordinals $\alpha_1 < \alpha_2 < \alpha_3 < \cdots$ has a countable least upper bound, or *limit*, obtained by collecting all the members of $\alpha_1, \alpha_2, \alpha_3, \ldots$. (To see why this collection is countable, recall how we packed countably many countable sets into Hilbert's hotel, in Section 1.2.) It follows that *the set of all countable ordinals* (the axioms of set theory guarantee that such a set exists) *is uncountable*. Because if there were only countably many countable ordinals, the successor of their least upper bound would be a countable ordinal greater than all countable ordinals—which is a contradiction.

REMARKS ON NOTATION AND REALIZATION

It should now be clear why we did not use the symbol $\aleph_0$, already used for the first transfinite cardinal, for the first transfinite ordinal. The concept of ordinal is a refinement of the concept of cardinal, because the ordinal number of a set depends on the ordering of the set, not simply on its cardinality. Indeed, we have just seen that there are uncountably many countable ordinal numbers—all corresponding to well-orderings of a set of cardinality $\aleph_0$.

This means, unfortunately, that there is no complete system of names for countable ordinals, as we normally understand the concept of "name." In practice, "names" are strings of letters in some finite alphabet, and there is only a countable infinity of such strings. We can list them by listing the finitely many one-letter strings first, then the finitely many two-letter strings, then the finitely many three-letter strings, and so on. Thus, if one wants to name all countable ordinals uniformly, one has to take the more theoretical option of letting each ordinal (generally, an infinite set) be its own name.

Fortunately, in this book we need names mainly for ordinals up to ε_0, and a nice uniform system of such names exists. We describe this system more fully in Section 2.6.

Perhaps the most concrete way to realize countable ordinals is by sets of rational numbers, with the natural ordering. Indeed, we can stick to sets of rational numbers in the unit interval $(0,1)$. The ordinal ω, for example, is the ordinal number of the set $\left\{\frac{1}{2}, \frac{3}{4}, \frac{7}{8}, \ldots\right\}$. Figure 2.1 shows a picture of this set of rationals, marked by vertical lines whose height decreases towards zero as we approach ω.

If an ordinal α is representable in this way, so is $\alpha + 1$: first map the set of rationals representing α into $(0, 1/2)$ by dividing each member by 2; then insert the point $1/2$ to represent $\alpha + 1$. And if ordinals $\alpha_1 < \alpha_2 < \alpha_3 < \cdots$ are representable, then so is their least upper bound: map the rationals representing α_1 into $(0, 1/2)$, the rationals representing α_2 into $(1/2, 3/4)$, the rationals representing α_3 into $(3/4, 7/8)$, ..., and then insert the point 1 to realize the least upper bound of $\alpha_1, \alpha_2, \alpha_3, \ldots$. Finally, divide by 2 so that all the rationals used lie in $(0,1)$.

Figure 2.1. A set with ordinal number ω.

Figure 2.2. A set with ordinal number ω^2.

As an example, Figure 2.2 shows what ω^2 looks like when represented by a set of rationals.

Sets of rationals represent every ordinal obtained by taking successors and countable least upper bounds, and it is clear that every countable ordinal arises in this way. (If not, consider the least countable ordinal *not* obtainable: it is either a limit or a successor, so we get a contradiction.)

2.3 The Axiom of Choice

> ... even for an infinite totality of sets there are always mappings that associate with every set one of its elements.
> —Ernst Zermelo (1904), p. 516.

The collection of countable ordinals, ordered by size, is well ordered. This can be seen by considering any set S of countable ordinals, and choosing a member α of S. Either α is the smallest member of S, or else there are members of S less than α, which are therefore members of α. Among the latter members of S there is a least, because α is well-ordered. Thus, in either case, the set S of countable ordinals has a least member.

It follows that the set of countable ordinals is itself an ordinal number, which Cantor called ω_1. Since it has uncountably many members, ω_1 is necessarily an uncountable ordinal. Indeed, ω_1 is the *least uncountable ordinal*, since any ordinal less than it is countable. Thus the countable ordinals make up the smallest possible uncountable set, whose cardinality was called $\aleph_1$ by Cantor. The idea that produces the uncountable set $\aleph_1$ contrasts strongly with our first idea for producing an uncountable set— taking all $2^{\aleph_0}$ subsets of the positive integers. The cardinal number $\aleph_1$ is well-ordered by construction, whereas there is no way (short of adding a new axiom to ZF) to well-order a set with $2^{\aleph_0}$ members. And the size of $\aleph_1$ is the next size after $\aleph_0$, whereas the size of $2^{\aleph_0}$ is an almost complete mystery.

It is not even clear that $\aleph_1$ is less than $2^{\aleph_0}$, because they may not even be comparable! The inequality $\aleph_1 \leq 2^{\aleph_0}$ means that there is a one-to-one correspondence between a set of size $\aleph_1$ (say, the countable ordinals) and a subset of some set of size $2^{\aleph_0}$ (say, the set of positive integer functions).

Our scheme for associating countable ordinals with positive integer functions seems tailor-made to prove this fact, but there is a catch.

We started with the following labelling of functions by ordinals—

$$f_1(n) = n$$
$$f_2(n) = n^2$$
$$f_3(n) = n^3$$
$$\vdots$$
$$f_\omega(n) = n^n$$
$$f_{\omega+1}(n) = n^{n+1}$$
$$f_{\omega+2}(n) = n^{n+2}$$
$$\vdots$$
$$f_{\omega \cdot 2}(n) = n^{n^2}$$
$$\vdots$$

—and we can continue happily to $f_{\varepsilon_0}(n)$, and probably further. But exactly how do we define $f_\alpha(n)$, where α is an arbitrary countable ordinal, assuming that $f_\beta(n)$ has been defined for all ordinals $\beta < \alpha$?

There is no problem when α is a *successor* ordinal, that is, when $\alpha = \beta + 1$ for some β. In this case we let $f_\alpha(n) = n f_\beta(n)$. But when α is a *limit* ordinal, that is, not a successor, we have to choose a sequence

$$\beta_1 < \beta_2 < \beta_3 < \cdots$$

of ordinals with least upper bound α, and let $f_\alpha(n)$ be a function that grows faster than any member of the sequence

$$f_{\beta_1}(n), \quad f_{\beta_2}(n), \quad f_{\beta_3}(n), \quad \ldots,$$

using du Bois-Reymond's theorem. The resulting function f_α depends on the choice of sequence $\beta_1, \beta_2, \beta_3, \ldots$, even in the case $\alpha = \omega$.

It may seem that there is always a natural choice for the sequence with limit α, and indeed the famous English mathematician G. H. Hardy proposed exactly this method for constructing a set of $\aleph_1$ positive integer functions in 1904. Yet there is no explicit rule for choosing an increasing sequence with limit α that can be proved to work for every countable limit ordinal α. This quandary, and others where it is necessary to make infinitely many choices in the absence of any apparent rule, convinced mathematicians that set theory needed a new axiom: the so-called *axiom of choice*.

The formal statement of the axiom of choice is quite simple:

AXIOM OF CHOICE. *For any set X of nonempty sets S, there is a function* choose(*S*) *(a "choice function for X") such that* choose(*S*) *is in S for each set S in X.*

Some classical applications of the axiom of choice are given in Section 2.9. Here is how we use the axiom of choice to get out of our present quandary. For each countable limit ordinal α, let x_α be the set of *all* increasing sequences with limit α, and then let X be the set of all such x_α. A choice function F for X gives an increasing sequence $F(x_\alpha)$, with limit α, for each countable limit ordinal α. F is hardly a "rule" for finding a sequence with limit α. Rather, it is what one resorts to when no rule is available.

THE AXIOM OF CHOICE AND WELL-ORDERING

To save space we now abbreviate "the axiom of choice" by AC. AC is involved in many questions involving uncountable sets. Perhaps the simplest such question is the following: how does one choose a number from each nonempty set of real numbers? Any rule that comes to mind—such as choosing the least member—is easily seen to break down, as we mentioned in Section 2.2. In fact, only AC guarantees that a choice function for sets of real numbers even exists.

But AC is more than just a last resort when one wants to make infinitely many choices. AC literally brings order to the mysterious world of uncountable sets, because it implies that *every set can be well-ordered.* This was proved by Ernst Zermelo in 1904. In particular, AC implies that the set of real numbers can be well-ordered. Naturally, there is no "rule" for this well-ordering, but if we imagine being "given" the well-ordering, the real numbers become almost as easy to handle as a countable set. In particular, we can immediately solve the problem of choosing an element from each nonempty set of reals; namely, choose its "least" element in the sense of the well-ordering.

A similar argument shows that a choice function can be defined for the subsets of any well-ordered set. Hence AC is in fact *equivalent* to the assertion that every set can be well-ordered, the so-called *well-ordering theorem.*

It follows from the well-ordering theorem that *cardinal numbers can be represented by particular ordinals.* For example, $\aleph_0$ can be represented by ω, $\aleph_1$ by ω_1, and in general the cardinal number of any set S can be represented by the least ordinal number of well-orderings of S. It follows in turn that any two cardinal numbers are *comparable*: that is, one is less than or equal to the other. We compare cardinal numbers by comparing the ordinals that represent them. This is one of the ways in which AC brings order into the world of uncountable sets.

2.4 The Continuum Hypothesis

> $\aleph_1$ is the set of countable ordinals and this is merely a special and the simplest way of generating a higher cardinal. The [continuum] is, by contrast, generated by a totally new and more powerful principle, namely the Power Set Axiom.
>
> —Paul Cohen (1966), p. 151.

It would be nice if the two basic methods for producing an uncountable set—taking all $2^{\aleph_0}$ subsets of the set of size $\aleph_0$, and taking all $\aleph_1$ ordinals of size $\aleph_0$—led to the same cardinality. But it is not clear that they do. The problem of deciding whether $2^{\aleph_0} = \aleph_1$ is called the *continuum problem*, and it has been open since Cantor first raised the question in this form in 1883.[1] The claim that $2^{\aleph_0} = \aleph_1$ is called the *continuum hypothesis* (CH).

The 20th century brought many dramatic advances in our understanding of what is provable—some of which we will study later in this book—but the continuum problem remains a mystery. It seems to lie at the very limit of the human capacity to understand the infinite. Nevertheless, there have been two breakthroughs that at least help to explain why the continuum problem is difficult.

In 1938, the Austrian mathematician Kurt Gödel showed that CH is *consistent* with the standard axioms of set theory (the ZF axioms, listed in Section 1.9). That is, if the standard axioms are consistent, then they and CH *together* do not lead to a contradiction. However, this does not mean that CH can be proved from the standard axioms, either. Gödel also showed, a few years earlier, that the standard axiom system is *incomplete:* assuming that the standard axioms are consistent, many sentences can be neither proved nor disproved. It turned out that CH is one of them, though not by Gödel's original argument for incompleteness.

The unprovability of CH was shown in 1963 by the American mathematician Paul Cohen, whom we quoted at the beginning of this section. The quote is from a book that Cohen wrote in 1966, expounding his and Gödel's results. It is preceded by the speculation that

> a point of view which the author feels may eventually come to be accepted is that CH is *obviously* false.

---

[1]Since $\aleph_1$ is a well-ordered set, the equation $2^{\aleph_0} = \aleph_1$ implies that the real numbers can be well-ordered. This alone is a subtle question, as we have already seen. Cantor's first formulation of the continuum problem, in 1878, avoided having to decide whether the real numbers can be well-ordered. He simply asked whether each infinite set of real numbers can be put in one-to-one correspondence with either $\aleph_0$ or $2^{\aleph_0}$.

WARNING ABOUT UNPROVABLE SENTENCES

It should be stressed that *unprovability in* ZF (or in any other system with basic logic) *can only be demonstrated under the assumption that the system is consistent.* This is because, by basic logic, *all sentences are deducible from an inconsistency.* We shall give occasional reminders of this point, but it would be tedious to mention it on every occasion when sentences are claimed to be unprovable.

The point of this warning is that consistency of ZF (or of any reasonably strong axiom system) really *is* an assumption. We shall see later why this is so.

Gödel also believed that CH is false, and he conjectured that $2^{\aleph_0}$ equals $\aleph_2$, the cardinality of the set of all ordinals of cardinality $\aleph_1$.

However, no convincing new axioms that settle CH have yet been proposed. It is conceivable that CH cannot be settled by any axioms more plausible than itself. But how on earth would we prove *that?*

The Continuum Hypothesis and Rates of Growth

The continuum hypothesis gives a partial answer to the questions 2 and 3 that were raised in Section 2.1: can we "exhibit" a sequence of integer functions,

$$f_0, f_1, \ldots, f_\beta, \ldots, \quad \text{where each } \beta \text{ is a countable ordinal,}$$

such that each f_β grows faster than any of its predecessors, and such that the growth rate of any given integer function g is exceeded by that of some f_β? We cannot exactly "exhibit" such a sequence, but CH implies that *there exists* a well-ordered sequence of length ω_1 with the required property.

We know from Section 1.7 that there are $2^{\aleph_0}$ positive integer functions, so CH implies that all such functions may be arranged in a sequence

$$f_0, f_1, \ldots, f_\alpha, \ldots, \quad \text{where } \alpha \text{ runs through all the countable ordinals.}$$

It then suffices to run through this sequence, picking out a suitable functions f_β that grow faster than all their predecessors.

To be precise, first pick f_0. After that, pick each f_β that grows faster than all functions up to f_α, where α is the least upper bound of the indices of functions previously picked. There are only countably many functions up to the index α in the original sequence, hence there is a function h that grows faster than them all, by du Bois-Reymond's theorem. The first such h therefore occurs (and gets picked) as some f_β, with $\beta > \alpha$, since the sequence $f_0, f_1, \ldots, f_\alpha, \ldots$, includes *all* positive integer functions. Consequently, new functions f_β will be picked for β beyond any index $\alpha < \omega_1$ in the original sequence. And, by construction, the growth rate of each f_β exceeds that of any integer function preceding f_α in the sequence of all integer functions. Thus the growth rate of any given integer function g is exceeded by that of some f_β.

A sequence of functions such as the f_β is called a *scale*. It is like a ruler (with a sequence of ω_1 marks on it) that we can use to measure the growth rate of any positive integer function. The existence of scales was first conjectured by the French mathematician Émile Borel, in his book *Leçons sur la théorie des fonctions* (Lectures on the Theory of Functions) of 1898, p. 118, without appealing to CH. Borel relied on du Bois-Reymond's diagonal argument, but also on a vague belief that by "transfinite repetition" of diagonalization one could overtake the growth rate of any given integer function.

2.5 Induction

> The quantity $a + (b + e)$ is the successor of the quantity $a + b \ldots$
> —Hermann Grassmann (1861), §2, Section 16.

We can give a precise account of ordinal symbols such as $\omega + \omega$, ω^2, and ω^ω by defining the sum, product, and exponential functions for countable ordinals. Our definitions extend the so-called *inductive* definitions of these functions for finite ordinals to ω and beyond, using the fact that every ordinal is either 0, a successor, or the least upper bound of its predecessors.

Inductive definitions of basic functions of arithmetic were first proposed by the German mathematician Hermann Grassmann in 1861. Grassmann recognized that the key property of the natural numbers 0, 1, 2, 3, 4, 5, ... is that they originate from 0 by repeating the operation of adding 1 (which he denoted by e). This *inductive property* of natural numbers allows a function $f(n)$ to be defined for all natural numbers n by first giving the value $f(0)$, then giving $f(n + 1)$ in terms of the value $f(n)$ already defined.

Of course, when we say "in terms of $f(n)$" we mean that $f(n + 1)$ should be computed from $f(n)$ using some operation already known. In the beginning, the only "known" operation is adding 1—the *successor function*—so all we can do to $f(n)$ is add 1. This is enough to define the *sum function* $f(n) = m + n$, for any fixed number m, by the following pair of equations:

$$m + 0 = m,$$
$$m + (n + 1) = (m + n) + 1.$$

The first equation, which declares that $f(0) = m$, is called the *base step* of the definition. The second equation, which declares that $f(n + 1) = f(n) + 1$ (the successor of $f(n)$) is called the *induction step*.

Now that we have defined the operation of adding n, for any natural number n, we can use the operation to define the *product function* $g(n) = m \cdot n$ by the following two equations:

$$m \cdot 0 = 0,$$
$$m \cdot (n + 1) = m \cdot n + m.$$

And then we define the *exponential function* m^n by the equations:

$$m^0 = 1,$$
$$m^{n+1} = m^n \cdot m.$$

The induction step in each of these definitions defines the function for all numbers that can be reached from 0 by the successor operation— the natural numbers, also known as the finite ordinals. To reach limit ordinals such as ω we need to insert a *least upper bound step* into each definition. The least upper bound step gives $f(\beta)$ in terms of the values $f(\gamma)$, for γ less than the limit ordinal β. The rule is simply that

$$f(\text{least upper bound of the } \gamma) = \text{least upper bound of the values } f(\gamma).$$

For the functions above we have, writing lub for least upper bound,

$$\alpha + \beta = \operatorname*{lub}_{\gamma < \beta}(\alpha + \gamma), \quad \alpha \cdot \beta = \operatorname*{lub}_{\gamma < \beta} \alpha \cdot \gamma, \quad \alpha^\beta = \operatorname*{lub}_{\gamma < \beta} \alpha^\gamma.$$

We now write α to denote any countable ordinal instead of m. The initial equations $m + 0 = m$, and so on, should also be written with α in place of m.

These definitions show that the sum, product, and exponential functions for ordinals naturally extend the corresponding functions on natural numbers. In particular, they show why we use the notation $\omega + 1$ for the successor of ω, why we write $\omega + \omega$ for the limit of the sequence $\omega + 1$, $\omega + 2, \omega + 3, \ldots$, why $\omega \cdot 2 = \omega + \omega$, and why we write ω^ω for the limit of the sequence $\omega, \omega^2, \omega^3, \ldots$.

Proof by Induction

Induction from n to $n+1$ in the domain of finite numbers is now replaced by the following: A statement $f(\alpha)$ concerning an ordinal number α is true for every α if $f(0)$ is true and if the truth of $f(\alpha)$ follows from the truth of $f(\xi)$ for all $\xi < \alpha$.

—Felix Hausdorff (1937/1957), p. 72.

Defining a function f by induction gives a unique value $f(\alpha)$ for each countable ordinal α. Proving this fact is an example of *proof by induction*— the natural accompaniment to definition by induction. For example, how do we know that the definition

$$m + 0 = m, \quad m + (n+1) = (m+n) + 1,$$

gives a unique value to $m+n$ for all natural number values of m and n? Think of "$m+n$ is unique for all m" as a proposition about the natural number n. This proposition is true for $n = 0$, because $m + 0$ has the unique value m. And if the proposition is true for a value $n = k$— that is, if $m + k$ has a unique value—then it is true for $n = k + 1$, since $m + (k+1)$ has the value $(m+k) + 1$, the unique successor of $m + k$. Thus the proposition is true for all n, because any n can be reached from 0 by repeating the successor operation.

The uniqueness of $m + n$ comes down to the uniqueness of successor.

It is not much harder to prove that $\alpha + \beta$ is unique for all ordinals α and β, thanks to the uniqueness of least upper bounds. Think of "$\alpha + \beta$ is unique for all α" as a proposition about the ordinal β. Again, this proposition is true for $\beta = 0$. Now we suppose that the proposition is true for all β less than some δ. It follows that the proposition is true for $\beta = \delta$—that is, $\alpha + \delta$ is unique—by the uniqueness of successor if δ is a successor and otherwise by the uniqueness of least upper bounds, because $\alpha + \delta = \text{lub}_{\beta < \delta}(\alpha + \beta)$. Thus $\alpha + \beta$ is unique for all β, because any ordinal β can be reached as either the successor or the least upper bound of smaller ordinals.

Thus the uniqueness of $\alpha + \beta$ (and, similarly, any function defined by transfinite induction) extends to all ordinals.

But while the *definitions* of these functions extend into the transfinite, not all of their properties do. For example, $m + n = n + m$ for any finite ordinals m and n. This was proved by Grassmann directly from the inductive definition. Yet it follows from the limit case in the definition of sum that

$$1 + \omega = \underset{i < \omega}{\text{lub}}(1 + i) = \omega,$$

so $1 + \omega \neq \omega + 1$. On the other hand, some useful properties of sum, product, and exponential that we know for natural numbers do extend

THE ASSOCIATIVE LAW FOR FINITE ORDINALS

Think of $l + (m + n) = (l + m) + n$ as a proposition about n.
This proposition is true for $n = 0, 1$ from the definition of sum.
And if $n = k + 1$ we have, assuming $l + (m + k) = (l + m) + k$,

$$l + (m + (k + 1)) = l + ((m + k) + 1) \quad \text{by definition of sum}$$
$$= (l + (m + k)) + 1 \quad \text{by induction hypothesis}$$
$$= (((l + m) + k) + 1) \quad \text{by induction hypothesis}$$
$$= (l + m) + (k + 1) \quad \text{by definition of sum.}$$

Thus the proposition is true for all finite n, by induction.

to ordinals. An example is the *associative law for addition*: $l + (m + n) = (l + m) + n$.

This proof is essentially the same as Grassmann's 1861 proof for finite ordinals. To extend his argument to all ordinals we need only deal with the limit case.

There are similar proofs for the *left distributive laws* (for multiplication and exponentiation) $\alpha(\beta + \gamma) = \alpha\beta + \alpha\gamma$ and $\alpha^{\beta+\gamma} = \alpha^\beta \cdot \alpha^\gamma$. I recom-

THE ASSOCIATIVE LAW FOR ALL ORDINALS

To prove $\alpha + (\beta + \gamma) = (\alpha + \beta) + \gamma$, think of this as a proposition about γ. The proposition is true for $\gamma = 0, 1$ (and all ordinals α and β), by the argument above.
Now suppose that $\alpha + (\beta + \gamma) = (\alpha + \beta) + \gamma$ holds for all γ less than some limit ordinal δ (and all α and β). Then the proposition also holds for $\gamma = \delta$ because

$$\alpha + (\beta + \delta) = \alpha + (\beta + \lub_{\gamma < \delta} \gamma)$$
$$= \lub_{\gamma < \delta}[\alpha + (\beta + \gamma)] \quad \text{by definition of sum}$$
$$= \lub_{\gamma < \delta}[(\alpha + \beta) + \gamma] \quad \text{by induction hypothesis}$$
$$= (\alpha + \beta) + \delta \quad \text{by definition of sum.}$$

Thus $\alpha + (\beta + \gamma) = (\alpha + \beta) + \gamma$ holds for all ordinals γ.

mend that readers try their hand at them, and also show that $(\beta + \gamma)\alpha$ is *not* always equal to $\beta\alpha + \gamma\alpha$.

2.6 CANTOR NORMAL FORM

The example $1 + \omega = \omega \neq \omega + 1$ shows that one has to be careful about the order of terms in a sum of countable ordinals.

The equation $1 + \omega = \omega$ shows that an ordinal gets absorbed when a "much larger" ordinal is added on its right. So we can omit ordinals in sums to the left of "much larger" ones. A similar thing happens with products. Since $\alpha^\beta \cdot \alpha^\gamma = \alpha^{\beta+\gamma}$ for ordinals α, β, and γ (as mentioned in the previous section), we have

$$\omega \cdot \omega^\omega = \omega^1 \cdot \omega^\omega = \omega^{1+\omega} = \omega^\omega.$$

Thus we can omit powers of ω in products to the left of "much larger" powers.

Omitting redundant terms in sums and products leads to a notation for ordinals known as the *Cantor normal form*. This notation is needed only for ordinals less than ε_0, that is, ordinals less than some member of the sequence

$$\omega, \quad \omega^\omega, \quad \omega^{\omega^\omega}, \quad \omega^{\omega^{\omega^\omega}}, \quad \ldots \ .$$

Such an ordinal α necessarily falls between two terms of this sequence, and example will suffice to show how we obtain its notation.

- Suppose that $\omega^\omega < \alpha < \omega^{\omega^\omega}$. It follows that α falls between two terms of the following sequence with limit ω^{ω^ω}:

$$\omega^\omega, \quad \omega^{\omega^2}, \quad \omega^{\omega^3}, \quad \ldots \ .$$

- Suppose $\omega^{\omega^2} < \alpha < \omega^{\omega^3}$. It follows that α falls between two terms of the following sequence with limit ω^{ω^3}:

$$\omega^{\omega^2}, \quad \omega^{\omega^2 \cdot 2}, \quad \omega^{\omega^2 \cdot 3}, \quad \ldots \ .$$

- Suppose that $\omega^{\omega^2 \cdot 7} < \alpha < \omega^{\omega^2 \cdot 8}$. It follows that α falls between two terms of the following sequence with limit $\omega^{\omega^2 \cdot 7 + \omega^2} = \omega^{\omega^2 \cdot 8}$:

$$\omega^{\omega^2 \cdot 7}, \quad \omega^{\omega^2 \cdot 7 + \omega}, \quad \omega^{\omega^2 \cdot 7 + \omega \cdot 2}, \quad \omega^{\omega^2 \cdot 7 + \omega \cdot 3}, \quad \ldots \ .$$

- Suppose that $\omega^{\omega^2 \cdot 7 + \omega \cdot 4} < \alpha < \omega^{\omega^2 \cdot 7 + \omega \cdot 5}$. It follows that α falls between two terms of the following sequence with limit $\omega^{\omega^2 \cdot 7 + \omega \cdot 4 + \omega} = \omega^{\omega^2 \cdot 7 + \omega \cdot 5}$:

$$\omega^{\omega^2 \cdot 7 + \omega \cdot 4}, \quad \omega^{\omega^2 \cdot 7 + \omega \cdot 4 + 1}, \quad \omega^{\omega^2 \cdot 7 + \omega \cdot 4 + 2}, \quad \ldots \; .$$

- Suppose that $\omega^{\omega^2 \cdot 7 + \omega \cdot 4 + 11} < \alpha < \omega^{\omega^2 \cdot 7 + \omega \cdot 4 + 12}$. This means that α is a term in the following sequence with limit $\omega^{\omega^2 \cdot 7 + \omega \cdot 4 + 11} + \omega = \omega^{\omega^2 \cdot 7 + \omega \cdot 4 + 12}$:

$$\omega^{\omega^2 \cdot 7 + \omega \cdot 4 + 11}, \quad \omega^{\omega^2 \cdot 7 + \omega \cdot 4 + 11} + 1, \quad \omega^{\omega^2 \cdot 7 + \omega \cdot 4 + 11} + 2, \quad \ldots \; .$$

So we could have, for example, $\alpha = \omega^{\omega^2 \cdot 7 + \omega \cdot 4 + 11} + 17$, which is the *Cantor normal form* for this particular α.

In general, each ordinal $\alpha < \varepsilon_0$ can be written in the form

$$\omega^{\alpha_1} \cdot n_1 + \omega^{\alpha_2} \cdot n_2 + \cdots + \omega^{\alpha_k} \cdot n_k,$$

where $\alpha_1 > \alpha_2 > \cdots > \alpha_k$ are ordinals less than α—call them the *exponents* of α—and $n_1, n_2, \ldots, n_k$ are positive integers. (This does not hold for $\alpha = \varepsilon_0$, because $\varepsilon_0 = \omega^{\varepsilon_0}$). Since $\alpha_1, \alpha_2, \ldots, \alpha_k$ are less than ε_0, they too can be expressed in this form, and so on. And since the exponents of each ordinal are less than it, this process of writing exponents as sums of powers of ω eventually terminates—otherwise we would have an infinite descending sequence of ordinals, contrary to well-ordering. The symbol ultimately obtained is called the *Cantor normal form of α*.

2.7 Goodstein's Theorem

A more elaborate example of an ordinal in Cantor normal form is

$$\alpha = \omega^{\omega^\omega + \omega} + \omega^{\omega^\omega} + \omega^\omega + \omega + 1.$$

Ordinary positive integers also have normal forms like the Cantor normal form, obtained by replacing ω by a fixed integer $b \geq 2$. With $b = 2$ we get the *base 2 Cantor normal form*, which stems from the fact that each positive integer n can be uniquely expressed as a sum of powers of 2. For example,

$$87 = 64 + 16 + 4 + 2 + 1 = 2^6 + 2^4 + 2^2 + 2^1 + 1.$$

The powers of 2 are found, uniquely, by first writing down the largest power of 2 not greater than the given number, then the largest power of 2 not greater than the given number minus the largest power of 2, and so on. This gives a series of exponents (for 87 they are 6, 4, 2, 1) which can also be written as sums of powers of 2, and so on. In this case we obtain, in the end,

$$87 = 2^{2^2+2^1} + 2^{2^2} + 2^2 + 2^1 + 1.$$

Thus 87 is a kind of "base 2 analogue" of the ordinal number α above.

There are similar "Cantor normal forms" for integers to base $b = 3$, 4, 5, and so on, because each positive integer is a unique sum of powers of b. We arrive at the general result by the same argument: subtract the largest power of b, and repeat. When $b > 2$ it may be possible to subtract the largest power of b up to $b - 1$ times, but this means only that there can be positive integer coefficients in the "base b Cantor normal form"— as in the Cantor normal form ordinals obtained by replacing b by ω. For example, in base 5,

$$87 = 5^2 \cdot 3 + 5^1 \cdot 2 + 2,$$

which corresponds to the ordinal $\omega^2 \cdot 3 + \omega \cdot 2 + 2$. When the base is b, each coefficient is at most $b - 1$.

In 1944, the English mathematician R. L. Goodstein noticed that this correspondence between ordinals and numbers to base b leads to a surprising property of positive integers. Namely, *the following process terminates at the number 0 for any positive integer m_1:*

GOODSTEIN PROCESS. *Given a positive integer m_1, write m_1 in base 2 Cantor normal form. Let n_1 be the number that results from replacing each 2 in the base 2 Cantor normal form of m_1 by 3, and let $m_2 = n_1 - 1$.*

Now write m_2 in base 3 Cantor normal form, replace each 3 by a 4 to get a new number n_2, and let $m_3 = n_2 - 1$.

Now write m_3 in base 4 Cantor normal form, replace each 4 by a 5 to get n_3, and let $m_4 = n_3 - 1$.

Now write m_4 in base 5 Cantor normal form, and so on.

Changing the base from b to $b + 1$ tends to create a much larger number, and subtracting 1 would seem to be of almost no help in reducing its size. But replacement of b by $b + 1$ can only take place when the Cantor normal form actually contains occurrences of b. It so happens that after a time (usually a very long time) all occurrences of b disappear. As we will see, this disappearance is due to the behavior of the Cantor normal form for ordinals, which controls the behavior of the base b normal form. Ultimately, the finiteness of the Goodstein process is due to the fact that any descending sequence of ordinals is finite.

It is hard to give helpful numerical examples of the Goodstein process, because there is a sudden leap from trivial to intractable as m_1 changes from 3 to 4. But let us try. We can at least use the case $m_1 = 3$ to illustrate

base	m	normal form of m	corresponding ordinal
2	m_1	$2+1$	$\omega + 1$
3	m_2	3	ω
4	m_3	3	3
5	m_4	2	2
6	m_5	1	1
7	m_6	0	0

Table 2.1. The Goodstein process for $m_1 = 3$.

the style of tabulation we will use in our attempt to describe the case $m_1 = 4$. The main thing to notice for $m_1 = 3$ is that, while the values of the normal form numbers do not always decrease, the corresponding ordinals always do. One of the disappointing trivial aspects of this example, shown in Table 2.1, is that the normal forms have only one term except for $m_1 = 2 + 1$.

Now we try the case $m_1 = 4 = 2^2$ in Table 2.2.

The step to m_2 is already a little tricky. First, we change each 2 to a 3 in m_1 to get $n_1 = 3^3$, then subtract 1, which gives (in base 3 normal form) $m_2 = 3^2 \cdot 2 + 3 \cdot 2 + 2$. Apart from such steps, where subtraction of 1 causes a drastic change in the shape of the normal form, there are long stretches where only obvious things happen, so we do a lot of skipping (indicated by dots). Even so, we can only roughly indicate what happens when the ordinals fall below the value $\omega^2 \cdot 2$. (According to Kirby and Paris (1982), the process reaches 0 at base $3 \cdot 2^{402653211} - 1$.[2])

Notice that we reach

ordinal $\omega^2 + \omega \cdot 23$ at base 47,
ordinal $\omega^2 + \omega \cdot 22$ at base $95 = 47 \cdot 2 + 1$,
ordinal $\omega^2 + \omega \cdot 21$ at base $191 = 95 \cdot 2 + 1$,

and so on—doubling the base and adding 1 each time. To get down to ω^2 we need to more than double the base another 20 times. After that, we drop to some ordinal of the form $\omega \cdot m + n$, and similarly reduce the positive integer m until we reach ω. The next step drops us to some finite (though enormous) ordinal k, at which point there is no symbol for the base in the Cantor normal form, so we can finally descend gracefully to 0, by $k - 1$ subtractions of 1.

The key to the termination of the Goodstein process (not only for $m_1 = 4$, but any value of m_1) is its correlation with a descending sequence of ordinals, which is necessarily finite by well-ordering. But why does

---

[2] Also, $402653211 = 3 \cdot 2^{27} + 27$. I learned this from two students at Mathcamp 2007, David Field and Jeremiah Siochi, who worked out this case of the Goodstein process.

base	m	normal form of m	corresponding ordinal
2	m_1	2^2	ω^ω
3	m_2	$3^2 \cdot 2 + 3 \cdot 2 + 2$	$\omega^2 \cdot 2 + \omega \cdot 2 + 2$
4	m_3	$4^2 \cdot 2 + 4 \cdot 2 + 1$	$\omega^2 \cdot 2 + \omega \cdot 2 + 1$
5	m_4	$5^2 \cdot 2 + 5 \cdot 2$	$\omega^2 \cdot 2 + \omega \cdot 2$
6	m_5	$6^2 \cdot 2 + 6 + 5$	$\omega^2 \cdot 2 + \omega + 5$
7	m_6	$7^2 \cdot 2 + 7 + 4$	$\omega^2 \cdot 2 + \omega + 4$
$\vdots$	$\vdots$	$\vdots$	$\vdots$
11	m_{10}	$11^2 \cdot 2 + 11$	$\omega^2 \cdot 2 + \omega$
12	m_{11}	$12^2 \cdot 2 + 11$	$\omega^2 \cdot 2 + 11$
13	m_{12}	$13^2 \cdot 2 + 10$	$\omega^2 \cdot 2 + 10$
$\vdots$	$\vdots$	$\vdots$	$\vdots$
23	m_{22}	$23^2 \cdot 2$	$\omega^2 \cdot 2$
24	m_{23}	$24^2 + 24 \cdot 23 + 23$	$\omega^2 + \omega \cdot 23 + 23$
25	m_{24}	$25^2 + 25 \cdot 23 + 22$	$\omega^2 + \omega \cdot 23 + 22$
$\vdots$	$\vdots$	$\vdots$	$\vdots$
47	m_{46}	$47^2 + 47 \cdot 23$	$\omega^2 + \omega \cdot 23$
48	m_{47}	$48^2 + 48 \cdot 22 + 47$	$\omega^2 + \omega \cdot 22 + 47$
49	m_{48}	$49^2 + 49 \cdot 22 + 46$	$\omega^2 + \omega \cdot 22 + 46$
$\vdots$	$\vdots$	$\vdots$	$\vdots$
95	m_{94}	$95^2 + 95 \cdot 22$	$\omega^2 + \omega \cdot 22$
96	m_{95}	$96^2 + 96 \cdot 21 + 95$	$\omega^2 + \omega \cdot 21 + 95$
97	m_{96}	$97^2 + 97 \cdot 21 + 94$	$\omega^2 + \omega \cdot 21 + 94$
$\vdots$	$\vdots$	$\vdots$	$\vdots$
191	m_{190}	$191^2 + 191 \cdot 21$	$\omega^2 + \omega \cdot 21$
$\vdots$	$\vdots$	$\vdots$	$\vdots$
383	m_{382}	$383^2 + 383 \cdot 20$	$\omega^2 + \omega \cdot 20$
$\vdots$	$\vdots$	$\vdots$	$\vdots$

Table 2.2. The Goodstein process for $m_1 = 4$.

each step in the process lead to a smaller ordinal? This lies in the nature of the Cantor normal form

$$f(\omega) = \omega^{\alpha_1} \cdot n_1 + \omega^{\alpha_2} \cdot n_2 + \cdots + \omega^{\alpha_k} \cdot n_k, \quad \text{where} \quad \alpha_1 > \alpha_2 > \cdots > \alpha_k.$$

When we go to the corresponding integer form $f(b)$ by replacing each ω by b, and subtract 1, what happens? The smallest-powered term in $f(b)$, $b^{\alpha_k} \cdot n_k$, is replaced by a sum of generally smaller powers, and any power b^{α_k} in the replacement has a smaller coefficient. Consequently, the term $\omega^{\alpha_k} \cdot n_k$ in the corresponding ordinal is replaced by a smaller sum of powers—and thus *we get a smaller ordinal*.

The fact that the Goodstein process terminates for any positive integer m_1 is called *Goodstein's theorem*. It is a theorem about positive integers and the basic operations on them: addition, multiplication, and exponentiation. Yet the interaction of these operations is so complex that the methods of ordinary arithmetic seem powerless to prove Goodstein's theorem. Apparently, we have to resort to the arithmetic of ordinals less than ε_0. This mysterious state of affairs will be confirmed and clarified later, when we show that the complexity of integer arithmetic itself is "measured" by ε_0.

2.8 Hercules and the Hydra

In Greek mythology, Hercules performed twelve "labors," or heroic tasks. The first of these was to kill the Nemean lion and bring back its skin; the second was to destroy the Lernaean hydra. Killing the lion was relatively

Figure 2.3. *Hercules and the Hydra*, by Antonio Pollaiuolo.

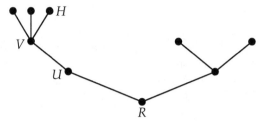

Figure 2.4. Example of a tree hydra.

easy—even though it had impenetrable skin—and Hercules succeeded by strangulation. After receiving a hint from Athena, he removed the skin by cutting it with the lion's own claws. The hydra proved more difficult, because it was a multi-headed beast with the ability to grow new heads in place of each head that was cut off. Hercules succeeded in this task only with the help of his nephew Iolaus, who used a fire brand to cauterize the wound each time a head was removed, thus preventing new heads from growing.

The painting of Hercules slaying the hydra by Antonio Pollaiuolo (Figure 2.3) tells the story with perhaps more drama than accuracy—how does he cut off its head? where is Iolaus?—though we must admit that Hercules is dressed to kill in his Nemean lion-skin suit.

A new variation on the hydra story was proposed by the English mathematicians Laurie Kirby and Jeff Paris in 1982. In their version, the hydra is shaped like a *tree*, such as the one in Figure 2.4. The tree has *root vertex R* and finitely many vertices connected to it by lines called *edges*. The sequence of edges, or *branch*, connecting the root vertex to any other vertex is unique. Each end vertex is a *head* of the hydra. Thus Figure 2.4 is a hydra with five heads. In this example, H is a head, and H is connected to R by the branch through the vertices V and U.

A head H is cut off by removing it, together with the edge connecting H to the neighboring vertex V. As in the original story, the hydra immediately grows new heads, but according to the following rules:

1. New heads grow only from the vertex U one edge below V on the path from U down to R.

2. If H is the nth head removed, n new copies of the decapitated portion of the hydra that remains above U grow from U.

Figure 2.5 shows the effect of the first decapitation, in the case where the head H removed is the one shown in Figure 2.4.

As this example shows, decapitation followed by new growth can easily increase the total number of heads. Nevertheless, *no matter how Hercules proceeds to cut off heads, eventually no heads remain.*

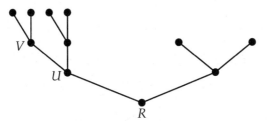

Figure 2.5. Tree hydra after head H is removed.

To prove this, we assign an ordinal $\alpha < \varepsilon_0$ to each tree, and show that each decapitation leads to a tree with smaller ordinal. The key to success is again the Cantor normal form.

The ordinal of a tree is obtained by assigning 0 to each head, then assigning the ordinal

$$\omega^{\alpha_1} + \omega^{\alpha_2} + \cdots + \omega^{\alpha_k}$$

to the vertex below the vertices with ordinals $\alpha_1, \alpha_2, \ldots, \alpha_k$. The ordinal assigned to R by this process is the ordinal of the tree. For example, Figure 2.6 shows the ordinals assigned to each vertex of the tree in Figure 2.4, with the tree itself assigned the ordinal $\omega^{\omega^3} + \omega^2$.

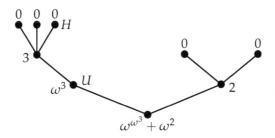

Figure 2.6. Tree hydra with ordinals assigned.

Figure 2.7 shows the ordinals after removal of the head H, assuming that H is the first head removed.

If H were the second head removed, two new subtrees would grow from the vertex U instead of one, and the ordinal assigned to U would be $\omega^2 \cdot 3$. But even if the ordinal assigned to U were $\omega^2 \cdot k$ (that is, if H were the $(k-1)$st head removed), the ordinal of the tree would be less than it was before removal of H. This is because $\omega^2 \cdot k < \omega^3$ for any finite ordinal k.

The reader can probably see that a similar argument allows us to conclude that the ordinal of the tree is *always* decreased when a head is

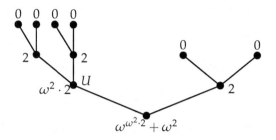

Figure 2.7. Ordinals after head H is removed.

removed, but we will not go into more detail. The point is, as usual, that any decreasing sequence of ordinals is finite. Therefore, after finitely many decapitations, no heads remain. Hercules can always destroy the hydra, with no cauterization necessary!

2.9 HISTORICAL BACKGROUND

"STRANGE" SETS AND FUNCTIONS

> ... from the point of view of logic, it is these strange functions that are most general; those that are met without being looked for no longer appear as more than a special case, and they have quite a little corner left them.
>
> —Henri Poincaré
> *Science and Method*, Chapter II, Section 5.

Cantor introduced ordinal numbers in 1872 in a study of *Fourier series*, the representation of functions as infinite sums of waves. This idea originated in the mid-18th century, when it was realized that the motion of a vibrating string is a sum of so-called "simple modes," each in the shape of the sine or cosine wave. Since the string can be set vibrating from a virtually arbitrary shape, it seems to follow that a virtually arbitrary function can be expressed as an infinite sum of sine or cosine functions. In 1808, Joseph Fourier developed an extensive theory of such sums—hence the name "Fourier series"—and gave specific examples, even for *discontinuous* functions.

One of Fourier's examples is the "square wave" shown in Figure 2.8, which takes the constant value $\pi/4$ over the interval $-\pi/2 < x < \pi/2$, and alternates this value with $-\pi/4$ over successive intervals of length π.

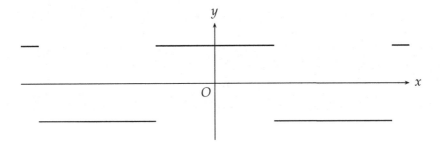

Figure 2.8. Square wave function.

Fourier showed that this function is the sum of the infinite series[3]

$$f(x) = \cos x - \frac{1}{3}\cos 3x + \frac{1}{5}\cos 5x - \frac{1}{7}\cos 7x + \cdots,$$

and hence it is the limit (as $n \to \infty$) of the continuous functions

$$f_n(x) = \cos x - \frac{1}{3}\cos 3x + \frac{1}{5}\cos 5x - \cdots + (-1)^n \frac{1}{2n+1}\cos(2n+1)x.$$

Figure 2.9 shows how well f_n approximates the square wave when $n = 20$.

Fourier's example shows that a function with points of discontinuity may nevertheless have a Fourier series. Cantor wanted to know "how badly" discontinuous a function could be, and still have a Fourier series. First, he needed to know how bad a set of potential points of discontinuity could be.

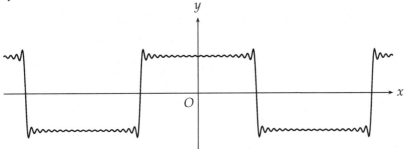

Figure 2.9. Approximation to the square wave.

---

[3]He also pointed out that, when $x = 0$, one gets the famous series

$$\frac{\pi}{4} = 1 - \frac{1}{3} + \frac{1}{5} - \frac{1}{7} + \cdots.$$

A case "worse" than Fourier's example is where the points of discontinuity have a limit point. This happens, for example, when there are points of discontinuity at $x = 1/2, 3/4, 7/8, 15/16, \ldots$, a set of points with limit $x = 1$. Worse still, there can be a "limit point of limit points," because we can insert an infinite sequence in front of each term in the sequence $1/2, 3/4, 7/8, 15/16, \ldots$. For example, insert

$$1/4, \ 3/8, \ 7/16, \ \ldots \quad \text{in front of } 1/2,$$
$$5/8, \ 11/16, \ 23/32, \ \ldots \quad \text{in front of } 3/4,$$
$$13/16, \ 27/32, \ 55/64, \ \ldots \quad \text{in front of } 7/8,$$
and so on.

Clearly, this insertion process—making any point of a set a limit point of a new set—may be repeated once, twice, $\ldots$, any finite number of times. So, if we take the *reverse* process of *removing all points that are not limit points*, the removal process may occur any finite number of times before the set is emptied. Then, by stringing together the results of one, two, $\ldots$ repetitions of the insertion process, we can create a set that is emptied only by *infinitely many* repetitions of the removal process.

And this is still not the worst that can happen. After infinitely many repetitions of the removal process we may be left with $1/2, 3/4, 7/8,$ $15/16, \ldots$, in which case we need one more removal of non-limit points to empty the set. This is how Cantor came to realize that it is necessary to be able to "count past infinity." It was the setting for his invention of transfinite ordinal numbers.

Once Cantor glimpsed the vast new world of sets and functions brought to light by this train of thought, the question of Fourier series seemed less important. Instead, given that the overwhelming majority of functions are not continuous, it became an interesting question to *classify* functions in a natural way. This question, too, leads to transfinite ordinals.

The road from analysis to transfinite ordinals became well-traveled around 1900, following the lead of the French mathematicians René-Louis Baire, Émile Borel, and Henri Lebesgue, all of whom came of age in the 1890s. Unlike their older compatriot Henri Poincaré, quoted at the beginning of this section, they welcomed the "strange" functions generated by the limit process.

They felt, since analysis is based on limits and continuous functions, that analysis should study *all* functions obtainable from continuous functions by taking limits. Baire suggested this approach to the function concept in 1898. From his point of view, many functions previously considered "strange" look quite simple and natural. Take, for

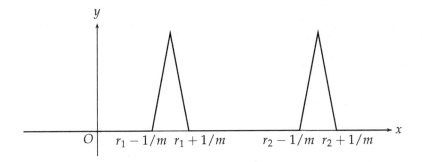

Figure 2.10. "Double-spike" function d_{2m} that has limit d_2 as $m \to \infty$.

example, the so-called *Dirichlet function* (as Baire himself did),

$$d(x) = \begin{cases} 1 \text{ if } x \text{ is rational} \\ 0 \text{ otherwise.} \end{cases}$$

The Dirichlet function is discontinuous at every point x. However, d is the limit of functions that are only "slightly" discontinuous. If we enumerate the rational numbers as $r_1, r_2, r_3, \ldots$, then $d(x)$ is the limit as $n \to \infty$ of the functions

$$d_n(x) = \begin{cases} 1 \text{ if } x = r_1, r_2, \ldots, r_n \\ 0 \text{ otherwise,} \end{cases}$$

each of which is discontinuous only at the finitely many points $r_1, r_2,$ $\ldots, r_n$. And each d_n, in turn, is the limit of a sequence of *continuous* functions. Figure 2.10 shows why, in the case $n = 2$.

The function $d_2(x)$, which has value 1 for $x = r_1, r_2$ and 0 otherwise, is the limit as $m \to \infty$ of "double-spike" functions $d_{2m}(x)$, with the properties

- $d_{2m}(x) = 1$ for $x = r_1, r_2$.

- $d_{2m}(x) = 0$ outside the intervals

$$\left[r_1 - \frac{1}{m}, r_1 + \frac{1}{m} \right] \text{ and } \left[r_2 - \frac{1}{m}, r_2 + \frac{1}{m} \right].$$

- $d_{2m}(x)$ increases continuously from 0 to 1 as x increases from $r_1 - \frac{1}{m}$ to r_1 and from $r_2 - \frac{1}{m}$ to r_2, and $d_{2m}(x)$ decreases continuously from 1 to 0 as x increases from r_1 to $r_1 + \frac{1}{m}$ and from r_2 to $r_2 + \frac{1}{m}$.

The "spikes" shrink to zero width as $m \to \infty$, leaving the values $d_2(x) = 1$ for $x = r_1, r_2$, and value 0 everywhere else. Similarly, d_n is the limit of continuous "n-tuple-spike" functions.

Thus the Dirichlet function is only two levels away from the continuous functions, in a hierarchy of functions that probably has infinitely many levels. In fact, Lebesgue showed in 1905 that the Baire hierarchy of functions has a level α for each countable ordinal α. Level zero consists of the continuous functions, and level $\alpha + 1$ consists of the limits of functions at level α. A diagonal argument shows that level $\alpha + 1$ includes functions *not* at level α. If λ is a limit ordinal, then the functions at level λ are all those at levels less than λ.

Meanwhile, starting in his *Leçons sur la théorie des fonctions* of 1898, Borel had been studying the problem of *measuring* sets of real numbers. The simplest sets of real numbers are the *open intervals*

$$(a, b) = \{x : a < x < b\}.$$

These, and the corresponding *closed intervals* $[a, b] = \{x : a \leq x \leq b\}$, have measure $b - a$ (also called their *length*). Open and closed intervals with the same endpoints have the same length because a point has length zero. Length can now be defined for many other sets of reals by the following rules:

- *Countable additivity:* If $S_1, S_2, S_3, \ldots$ are disjoint sets of length $l_1, l_2, l_3, \ldots$ respectively, then the length of $S_1 \cup S_2 \cup S_3 \cup \cdots$ is $l_1 + l_2 + l_3 + \cdots$.

- *Difference:* If sets S_1 and S_2 have lengths l_1 and l_2 respectively, and S_2 is a subset of S_1, then the length of the *set difference*

$$S_1 - S_2 = \{x : x \text{ is in } S_1 \text{ but not in } S_2\}$$

is $l_1 - l_2$ (provided l_1 is not ∞).

Countable additivity implies that the length of any countable set of points is $0 + 0 + 0 + \cdots = 0$, which agrees with the argument we used in Section 1.5 to show that there are more than countably many reals.

The two rules determine measure for a very large collection of sets, now known as the *Borel sets* (of reals). The Borel sets in fact are all sets obtainable from open intervals by countable unions and set difference. The countable unions need not be disjoint, because any countable union of Borel sets equals a disjoint countable union of sets that are also Borel. Likewise, difference can always be taken relative to the set $\mathbb{R}$ of all real numbers, because $\mathbb{R}$ is a countable union of open intervals.

Just as some "strange functions" turn up at low levels among the Baire functions, some "strange sets" turn up at low levels in the Borel sets. The set of rational numbers, for example, is a countable union of points, and a point is ℝ minus a countable union of open intervals. In Lebesgue's 1905 paper in which the Baire functions are shown to form a hierarchy, the Borel sets are shown to form a similar hierarchy with members at all countable ordinal levels. Lebesgue also showed that the Borel sets are precisely the sets S of reals for which the *characteristic function* of S,

$$\text{char}_S(x) = \begin{cases} 1 \text{ if } x \text{ is in } S, \\ 0 \text{ if } x \text{ is not in } S, \end{cases}$$

is a Baire function. We have already seen an example of this correspondence between Borel sets and Baire functions: the Dirichlet function is the characteristic function of the set of rational numbers.

THE AXIOM OF CHOICE

In a later article I shall discuss the law of thought that says that it is always possible to bring any *well-defined* set into the form of a *well-ordered* set—a law which seems to me fundamental and momentous and quite astonishing by reason of its general validity.

— Georg Cantor (1883),
in Ewald (1996), volume II, p. 886.

The axiom of choice concerns the ability to make infinitely many choices, which was often assumed to be possible in the early days of set theory. For example, Cantor in 1895 offered the following argument in support of the claim that *every infinite set has a countable subset*.

Suppose that X is an infinite set. Choose a member x_1 of X. Then $X - \{x_1\}$ is also infinite, so we can choose one of its members, x_2. Likewise, $X - \{x_1, x_2\}$ is infinite, so we can choose one of its members, x_3, ... and so on. Thus X has a countable subset $\{x_1, x_2, x_3, \ldots\}$.

Few people doubt that "and so on" makes sense here, even though no rule for choosing has been given. However, the idea of making infinitely many choices is harder to accept when uncountably many choices are required—as when Hardy assumed the possibility of choosing an increasing sequence $\beta_1 < \beta_2 < \beta_3 < \cdots$ with limit α for each countable limit ordinal α (Section 2.3). And it becomes a real article of faith that there is a function F that chooses a real number $F(X)$ from each nonempty set X of real numbers.[4]

---

[4]Notwithstanding the difficulty in imagining such a function, equivalent assumptions were made in analysis before the emergence of set theory. For example, in 1872, Cantor offered a proof that a function f is continuous if it is *sequentially continuous*. One says that f is sequentially continuous at $x = a$ if $f(a_n)$ tends to $f(a)$ for any sequence $a_1, a_2, \ldots$ that tends to a. The proof involves another unconscious use of the axiom of choice.

Examples like these brought Ernst Zermelo to the realization, in 1904, that an axiom of choice was needed. Set theorists and logicians experimented with various equivalent statements of the axiom, before settling on the statement given in Section 2.3: *for any set X of sets S, there is a function F such that F(S) belongs to S for each nonempty S.* Zermelo (1904) created a sensation by showing that *the axiom of choice implies that every set can be well ordered.* The existence of a well-ordering for any set had been proposed by Cantor in 1883, but only as a "law of thought" that he hoped to return to later. This did not happen. Instead, Zermelo's proof completely changed the complexion of set theory by reducing well-ordering to a more "obvious," or at least more simple, axiom.

With the help of the Zermelo-Fraenkel axioms described in Section 1.8, the proof of the well-ordering theorem itself can be made almost obvious. (Indeed, Zermelo's axioms were intended not only to avoid paradoxes, but also to clarify the proof of well-ordering.) Suppose we want to well-order a set X. This amounts to assigning ordinals $0, 1, 2, \ldots$ to members of X, until X is exhausted. In fact, that is basically the proof! I repeat: *assign ordinals to members of X until X is exhausted.* In a little more detail, we "assign" ordinals $0, 1, 2, \ldots, \alpha, \ldots$ to members of X by a function f defined inductively by

$$f(0) = \text{choose}(X),$$
$$f(\beta) = \text{choose}(X - \{\text{values } f(\alpha) \text{ for } \alpha < \beta\}),$$

where $\text{choose}(S)$ is a choice function for the subsets S of X. Then X is "exhausted" because, if β is the least upper bound of the ordinals α for which $f(\alpha)$ is defined, then $X - \{\text{values } f(\alpha) \text{ for } \alpha < \beta\}$ is necessarily empty.

Today, most mathematicians have embraced the axiom of choice because of the order and simplicity it brings to mathematics in general. For example, the theorems that every vector space has a basis and every field has an algebraic closure hold only by virtue of the axiom of choice. Likewise, for the theorem that every sequentially continuous function is continuous. However, there are special places where the axiom of choice actually brings *dis*order. One is the theory of measure.

As mentioned in the previous subsection, Borel (1898) extended the concept of length from intervals of the line to the vast collection of Borel sets. In 1902, Lebesgue extended the concept as far as it can go, to what are called the *Lebesgue measurable* sets. In one sense, the extension is small, because every Lebesgue measurable set differs from a Borel set by a set of measure zero. On the other hand, the Borel sets are only as numerous as the real numbers, whereas the Lebesgue measurable sets are as numerous as the *sets* of real numbers. Depending on what one is willing to assume beyond ZF, Lebesgue measure may in fact extend to all sets of real numbers. But if the axiom of choice holds there are sets of reals that are *not* Lebesgue measurable.

This was discovered by the Italian mathematician Giuseppe Vitali in 1905, and it led to doubts about the axiom of choice on the part of Borel, Baire, Lebesgue, and some other distinguished analysts of the past. (For more on the crisis in analysis precipitated by the axiom of choice, see Moore (1982).)

VITALI'S NONMEASURABLE SET

This set is best viewed as a subset X of the circle with the following two properties:

- When X is rotated by an angle $2\pi r$, where r is rational, the resulting set X_r is disjoint from X or equal to X.

- The (infinitely many) sets X_r fill the circle.

It follows from these properties that X cannot have measure $m > 0$. If it did, each rotated set X_r would then have the same measure m, and the measure of the circle would be $m + m + m + \cdots = \infty$ by countable additivity. But X likewise cannot have measure zero, otherwise the measure of the circle would be $0 + 0 + 0 + \cdots = 0$. Thus a set with these two properties is not measurable. We find such a set with the help of the axiom of choice. The simplest way is to view the circle as the set of angles from 0 to 2π and to partition it into sets of the form

$$S_\theta = \{\text{angles } \theta + 2\pi r, \text{ where } r \text{ is rational}\}.$$

Then a suitable X is obtained by choosing one member from each distinct S_θ. (Beware of simply choosing θ from S_θ! The sets S_θ are the *same* for infinitely many values of θ, and just one of these values must be chosen to go in X.)

Vitali's result was the first in a series of increasingly counterintuitive results on measure obtained by use of the axiom of choice, culminating in the so-called "paradox" of Banach and Tarski (1924). The latter result is a decomposition of the unit ball into finitely many sets which, when moved to different positions, fill *two* solid balls. The sets into which the ball is decomposed exist only by virtue of the axiom of choice. They cannot be "measured" even by a *finitely* additive measure, as this would imply $1 = 2$. There is an excellent modern account of the Banach-Tarski paradox in Wagon (1993).

Thus, there is a downside to using the axiom of choice. If one is willing to abandon the axiom, then in fact it is consistent to assume that all sets of real numbers are Lebesgue measurable (assuming the consistency of ZF plus an "inaccessible cardinal"—see Chapter 7). This astonishing result was proved by the American mathematician Robert Solovay in the 1960s. It was part of a new wave of set theory that revealed just how deeply mysterious the real numbers are.

The Continuum Hypothesis

> From this theorem it would follow at once that the continuum has the next cardinal number beyond that of the countable set; the proof of this theorem would, therefore, form a new bridge between the countable set and the continuum.
>
> —David Hilbert, Address on *Mathematical Problems*, 1900, in Ewald (1996), volume II, p. 1103.

After Cantor discovered that the set $\mathbb{R}$ of real numbers is uncountable, it soon occurred to him to ask: is $\mathbb{R}$ the "next largest" infinite set? In other words, is every set of real numbers either countable or of the same cardinality as $\mathbb{R}$? This conjecture was his first version of the continuum hypothesis, proposed in 1878: *every uncountable set of reals has the same cardinality.* Then, in 1883, Cantor discovered the smallest uncountable set, the set of countable ordinals, and proposed the second version: $\mathbb{R}$ *has the same cardinality as the set of countable ordinals.* In symbols: $2^{\aleph_0} = \aleph_1$.

The latter version of the continuum hypothesis (CH) is much stronger, because it implies that $\mathbb{R}$ can be well-ordered. As we have seen, Cantor in 1883 believed that any set can be well-ordered, but he was not yet aware of the need for the axiom of choice.

Until he discovered the general diagonal argument, in 1891, Cantor did not realize that there were sets larger than $\mathbb{R}$. So he understandably viewed CH as the ultimate question in set theory. It remained on top even after Cantor's discovery of higher infinities. In 1900, when the German mathematician David Hilbert proposed 23 problems for the mathematicians of the 20th century, CH was number one on his list. Zermelo's proof of the well-ordering theorem, in 1904, eased the concern about well-ordering $\mathbb{R}$ to some extent, but it did not seem to help settle the continuum problem. Nor did the detailed investigation and classification of sets of real numbers initiated by Baire, Borel, and Lebesgue. Finally, Hilbert himself put forward an idea in a 1926 paper entitled *Über das Unendliche* ("On the Infinite").

The idea of Hilbert (1926) was to correlate countable ordinals with positive integer functions (which, as we saw in Section 1.7, are in one-to-one correspondence with the real numbers). However, he did not correlate ordinals with functions but rather with *formulas describing functions* in a certain language. The language itself is not clearly explained; functions are defined in terms of ordinals which are defined in terms of functions, in seemingly circular fashion. So Hilbert's "proof" is unconvincing. Nevertheless, it contains the germ of a much better idea, used successfully by Kurt Gödel in 1938 (see Gödel (1939) or Cohen (1966), p. 87–99, for details).

Gödel's idea is to build a *model of set theory* consisting of the smallest collection L of sets that includes all the ordinals and satisfies the Zermelo-Fraenkel axioms for sets enumerated in Section 1.8. Thus, L includes the empty set; along with any two members X and Y, L includes the unordered pair $\{X, Y\}$; and so on. However, in a faint echo of Hilbert, the only *subsets* in L of a set X in L are those *definable* in a certain language. The key difference between Hilbert and Gödel is that Gödel avoids circularity by *not* attempting to name the ordinals. Instead, he takes the ordinals as given and defines sets in terms of previously-defined sets (starting with the empty set) in ordinal-numbered stages. L is called the class of *constructible* sets.

L is the union of sets L_α, for all ordinals α: L_0 is empty, $L_{\alpha+1}$ consists of all subsets of L_α definable in the language of ZF with names for members of L_α, and (for limit β) L_β is the union of all L_α for $\alpha < \beta$. The result of this construction is the following.

- L satisfies all the Zermelo-Fraenkel axioms (ZF). Also, L is well-ordered, via a well-ordering of definitions.

  This is plausible because there are only countably many formulas in the language of ZF (by the remark on listing strings of letters in Section 2.2), hence ZF formulas can be well-ordered. Admittedly, the ZF formulas can get ordinals (and other sets in L_β) plugged into them to make definitions of sets in L_α, but the inductive construction of L_α ensures that the resulting definitions can also be well-ordered. Thus each set in L is well-ordered by the well-ordering of formulas. More subtly, the well-ordering is *itself* a set in L. Because of this, L satisfies the well-ordering theorem and hence the axiom of choice.

- L also satisfies CH. This is less obvious. It depends on techniques of mathematical logic which show that any set of integers in L occurs in some L_α, where α is countable. There are only $\aleph_0$ sets in each such L_α, because of the countability of ZF formulas, hence $\aleph_1$ sets altogether.

These properties of L do *not* prove that CH is true for sets in general. But they show that *CH cannot be disproved*, because L satisfies both ZF and CH. More precisely, if sets in general satisfy ZF, then the particular sets in L satisfy ZF plus CH. Or again, *if ZF is consistent, then so is ZF plus CH*. This is how Gödel came up with his proof, mentioned in Section 2.4, that CH cannot be disproved by the usual axioms of set theory.

Cohen's proof that CH cannot be proved, also mentioned in Section 2.4, involves a model of ZF in which CH is false. Cohen built such a model with the help of his brilliant innovation called *forcing*, that allows the construction of extremely varied models of ZF. To build a model in which CH is false he first cut down Gödel's L, which includes all the ordinals and hence is highly uncountable, to a *countable* model M. This can be done, essentially, because there are only countably many formulas in the language of ZF. Since M includes only countably many sets of integers, there is room for many new sets of integers $S_1, S_2, \ldots$ to be put back into M. When this is done very cleverly, one can obtain a model $M(S_1, S_2, \ldots)$ of ZF (and also AC) that does *not* satisfy CH. Thus *the continuum hypothesis cannot be proved* from the ZF axioms, even when AC is added (unless ZF itself is inconsistent).

Forcing

The technique of forcing is too technical to describe here, but we cannot ignore how important it has been. It led to most of the 1960s revelations in set theory, such as Solovay's theorem on the consistency of assuming that all sets of real numbers are measurable. Forcing completely changed the face of set theory by showing that *most of the interesting unproved sentences of ZF can be neither proved nor disproved without new axioms*.

Solovay's result is one of many showing that it is consistent to assume that AC is false, assuming that ZF is consistent. Thus, while AC cannot be disproved from the ZF axioms (by Gödel's result about L), it cannot be proved either. The undecidable sentences of ZF include not only the full axiom of choice, but also many special cases of it that we touched on earlier in this chapter. In particular, forcing can be used to obtain models of ZF with the following properties, which are therefore not disprovable in ZF:

- There is an infinite set with no countable subset.

- There is a sequentially continuous function that is not continuous.

- $\mathbb{R}$ is a countable union of countable sets.

These properties all contradict AC, so they are not provable in ZF either.

Figure 2.11. Sunset or sunrise?

At a conference in memory of Paul Cohen at Stanford on September 29, 2007, Hugh Woodin described Cohen's discovery by saying: *It is really the end of set theory, or else really the beginning—and we do not yet know which.* Woodin then displayed a photo of an alien landscape (a NASA photo of Mars, Figure 2.11) in which the sun appears to set, or rise—and one cannot tell which.

It may be that we can never understand the continuum; it may also be that we *can* understand the continuum, but only by new axioms (Woodin takes the optimistic view). Be that as it may, the two great ideas of Cantor—the diagonal argument and the ordinals—have thrown new light on mathematics even at the level of positive integers. There too, the usual axioms are not sufficient to settle all questions. We turn to these developments in the remaining chapters of the book.

- CHAPTER 3 -

COMPUTABILITY AND PROOF

PREVIEW

The concept of computability emerged in mathematics from the concept of *proof*. Around 1900, difficulties in the foundations of mathematics led to the demand for a *formal* concept of proof, in which theorems are derived from axioms by completely explicit rules of inference. The rules are supposed to operate on strings of symbols without the need for conscious intervention, so a *formal system* is a process for proving theorems by computation.

This means that formal systems are themselves mathematical structures, about which theorems can be proved. The most famous theorem of this type is *Gödel's first incompleteness theorem*, stating that any consistent formal system for arithmetic fails to prove some true sentences. A remarkable (and shocking) consequence is Gödel's *second incompleteness theorem*, according to which a "sufficiently strong" formal system for arithmetic can express its own consistency, but cannot prove it (unless the system is inconsistent).

Gödel proved his theorems in 1930, unaware that a general definition of formal system, and a corresponding general incompleteness theorem, had already been discovered by Emil Post. The role of computability in logic did not become clear, or generally accepted, until it was rediscovered by Church and Turing in the 1930s.

The general definition of computability also makes it possible to speak of arbitrary *algorithms*, and to prove that no algorithm exists for solving certain problems. Among these *unsolvable* problems are the problem of deciding which computer programs ultimately halt, and of deciding which sentences of arithmetic or logic are true. The basis of all these results is surprisingly simple—a computable variant of the diagonal construction from Chapter 1.

3.1 Formal Systems

> ... the investigations of Weierstrass and others of the same school have shown
> that, even in the common topics of mathematical thought, much more detail
> is necessary than previous generations of mathematicians had anticipated.
> —Whitehead and Russell (1910), p. 3.

In the late 19th century several new branches of mathematics emerged
from problems in the foundations of algebra, geometry, and analysis.
The rise of new algebraic systems, non-Euclidean geometry, and with
them the need for new foundations of analysis, created the demand for
greater clarity in both the subject matter and methods of mathematics.
This led to:

1. symbolic logic, in which concepts of logic were expressed by symbols and deduction was reduced to the process of applying *rules of
   inference*,

2. set theory, in which all mathematical concepts were defined in terms
   of sets and the relations of membership and equality,

3. axiomatics, in which each branch of mathematics is defined by certain *axioms*, the logical consequences of which are its *theorems*.

Around 1900 these ideas were merged in the concept of a *formal system*, a symbolic language capable of expressing all mathematical concepts, together with a set of propositions (axioms) from which theorems
could be derived by specific rules of inference. The definitive formal system of the early 20th century was the *Principia Mathematica* of Whitehead
and Russell (1910–1913).

The main aims of *Principia Mathematica* were *rigor* and *completeness*.
The symbolic language, together with an explicit statement of all rules
of inference, allows theorems to be derived only if they are logical consequences of the axioms. It is impossible for unconscious assumptions
to sneak in by seeming "obvious". In fact, all deductions in the *Principia*
system can in principle be carried out *without knowing the meaning of the
symbols*, since the rules of inference are pure symbol manipulations. Such
deductions can be carried out by a machine, although this was not the intention of *Principia*, since suitable machines did not exist when it was
written. The intention was to ensure rigor by keeping out unconscious
assumptions, and in these terms *Principia* was a complete success.

As for completeness, the three massive volumes of *Principia* were a
"proof by intimidation" that all the mathematics then in existence was
deducible from the *Principia* axioms, but no more than that. It was not
actually known whether *Principia* was even *logically* complete, that is, capable of deriving all valid principles of logic. In 1929 Gödel proved its

logical completeness, but soon after that he proved its *mathematical* in-completeness. We are now getting ahead of our story, but the underlying reason for Gödel's incompleteness theorem can be stated here: the weakness of *Principia* (and all similar systems) is its very objectivity. Since *Principia* can be described with complete precision, *it is itself a mathematical structure*, which can be reasoned about. A simple but ingenious argument then shows that *Principia* cannot prove all facts about itself, and hence it is mathematically incomplete.

From Principia Mathematica to Post's Normal Systems

> Symbolic Logic may be said to be Mathematics become self-conscious.
>
> —Post (1941),
> in Davis (2004), p. 345.

In 1920 the 23-year-old American mathematician Emil L. Post entered the field of mathematical logic by proving the *completeness and consistency of propositional logic*. As its name suggests, propositional logic concerns *propositions*, which are denoted by variables $P, Q, R, \ldots$, and the logical relations between them. The internal structure of propositions is not analyzed, only the effect of *connecting* them by words such as "and," "or," and "not." Such words are called *connectives*, and today the connectives "and," "or," and "not" are denoted by symbols $\wedge$, $\vee$, and $\neg$ respectively. Thus $P \vee Q$ denotes "P or Q," and $(\neg P) \vee Q$ denotes "(not P) or Q." The latter is commonly abbreviated $P \Rightarrow Q$ because it is equivalent to "P implies Q."

Principia Mathematica gave certain axioms for propositional logic, such as $(P \vee P) \Rightarrow P$, and certain rules of inference such as the classical rule of *modus ponens*: from P and $P \Rightarrow Q$, infer Q. Post proved that all valid formulas of propositional logic follow from the axioms by means of these rules, so *Principia* is *complete* as far as propositional logic is concerned.

Post also showed that propositional logic is consistent, by introducing the now familiar device of *truth tables*. Truth tables assign to each axiom the value "true," and each rule of inference preserves the value "true," so all theorems have the value "true" and hence are true in the intuitive sense. This property of propositional logic—that all its theorems are true—is called *soundness*. The same idea also shows that propositional logic has the (possibly weaker) property of *consistency*. That is, it does not prove any proposition P together with its negation $\neg P$, since if one of these has the value "true" the other has the value "false."

Completeness and consistency together solve the *deducibility problem* for propositional logic: to give an algorithm that determines, for any given proposition, whether it is a theorem.

We now know that propositional logic is far easier than the full *Principia*. Indeed Post's results had already been obtained, though not published, by the German logician Paul Bernays in 1918. However, what is interesting is that Post went straight ahead, attempting to analyze *arbitrary* rules of inference, and hence to describe arbitrary formal systems. In doing so, he hit upon the concept of *computability*, because it turns out that Post's formal system generate *all* sets that can be computably listed. Post took a "rule of inference" to consist of a finite set of *premises*:

$$g_{11}P_{i_{11}}g_{12}P_{i_{12}} \cdots g_{1m_1}P_{i_{1m_1}}g_{1(m_1+1)}$$

$$g_{21}P_{i_{21}}g_{22}P_{i_{22}} \cdots g_{2m_2}P_{i_{2m_2}}g_{2(m_2+1)}$$

$$\cdots$$

$$g_{k1}P_{i_{k1}}g_{k2}P_{i_{k2}} \cdots g_{km_k}P_{i_{km_k}}g_{k(m_k+1)}$$

which together produce a *conclusion*

$$g_1P_{i_1}g_2P_{i_2} \cdots g_mP_{i_m}g_{m+1}.$$

The g_{ij} are certain specific symbols or strings of symbols, such as the $\Rightarrow$ symbol in modus ponens, and the P_{kl} are arbitrary strings (such as the P and Q in modus ponens). Each P_{kl} in the conclusion is in at least one of the premises. Such "rules" include all the rules of *Principia* and, Post thought, any other rules that draw conclusions from premises in a determinate way.

Thus *any formal system F is a finite set of axioms and rules of the above form*. A *proof* in F is a finite sequence of strings, called the *lines* of the proof, each of which follows from preceding lines by one of the rules of F. A *theorem* of F is the last line of a proof.

Solving the deducibility problem for general systems of rules amounts to understanding all of mathematics, a seemingly hopeless task. But Post made surprising progress in the beginning. By mid-1921 he had proved that the theorems of any formal system can be produced by what he called a *normal system*, consisting of a single axiom and rules of the simple form

$$gP \quad \text{produces} \quad Pg'.$$

In other words, any string beginning with the string g may be replaced by the string in which g is removed and g' is attached at the other end.[1]

---

[1] Most of Post's work was not published at the time, for reasons that are explained in Urquhart (2009). His reduction to normal form first appeared in Post (1943). A reduction of a typical system of symbolic logic to a normal system may be seen in the appendix of Rosenbloom (1950).

At first this looked like a great step forward. Apparently, *understanding all mathematics reduces to understanding normal systems*. But after failing to analyze some very simple normal systems (see below), Post realized that the situation was the opposite of what he had first thought: instead of simplifying *Principia*, he had merely distilled its complexity into a smaller system.

A TAG SYSTEM

Normal systems include an even simpler class of systems that Post called "tag" systems, in which each g' depends only on the initial letter of g and all g have the same length. One such system uses only the letters 0 and 1, and each g has length 3. If g begins with 0, then $g' = 00$, and if g begins with 1 then $g' = 1101$. The left-hand end of the string therefore advances by three places at each step, trying to "tag" the right-hand end which advances by two or four places. For example, here is what happens when the initial string is 1010:

$$1010$$
$$01101$$
$$0100$$
$$000$$
$$00$$
$$0$$

and then the string becomes empty. In all cases that Post tried, the result was either termination (as here) or periodicity, but it is still not known whether this is always the case.

Theorems can indeed be produced by simple rules, but only because *any* computation can be reduced to simple steps. Predicting the outcome of simple rules is no easier than deciding whether arbitrary sentences of mathematics are theorems.[2] This realization led to a reversal of Post's program.

The reverse program was easier than the one he had set himself initially, which was essentially the following:

1. Describe all possible formal systems.

---

[2] A nice example of the intractability of simple systems was given recently by De Mol (2008), when she showed that the notorious "$3n + 1$ problem" is equivalent to a tag system on three letters in which the strings g have length 2.

2. Simplify them.

3. Hence solve the deducibility problem for all of them.

In reducing complicated rules to simple ones, Post became convinced that, for *any* system generating strings of symbols, there is a normal system that generates the same strings. But *it is possible to list all normal systems*, since each consists of finitely many strings of symbols on a finite alphabet. Hence it is possible to list all systems for generating theorems. This invites an application of the diagonal argument, described below. The outcome is that *for certain formal systems the deducibility problem is unsolvable*.

After this dramatic change of direction Post saw the true path as follows:

1. Describe all possible formal systems.
2. Diagonalize against them.
3. Show that some of them have unsolvable deducibility problem.

And he also saw one step further—the incompleteness theorem—because:

4. No sound formal system obtains all answers to an unsolvable problem.

3.2 Post's Approach to Incompleteness

> ...the best I can say is that I would have *proved* Gödel's theorem in 1921—had I been Gödel.
>
> —Emil Post,
> postcard to Gödel, 29 October 1938
> in Gödel (2003), p. 169.

We deal with the last step of Post's program first, to dispel the myth that incompleteness is a difficult concept. Let us define an algorithmic problem, or simply *problem*, to be a computable list of questions: $P = \langle Q_1, Q_2, Q_3, \ldots \rangle$. For example, the problem of recognizing primes is the list

$$\langle \text{"Is 1 prime?"}, \text{"Is 2 prime?"}, \text{"Is 3 prime?"}, \ldots \rangle.$$

A problem is said to be *unsolvable* if the list of answers is not computable. The problem of recognizing primes is of course *solvable*.

Now suppose that an unsolvable $P = \langle Q_1, Q_2, Q_3, \ldots \rangle$ exists. Then no sound formal system F proves all true sentences of the form

$$\text{"The answer to } Q_i \text{ is } A_i \text{."},$$

since by systematically listing all the theorems of F we could compute a list of answers to problem P. A conceptually simple, though highly inefficient, way to list theorems is the following:

1. List all strings of symbols in the language of F (including the line separation symbol, whatever it is), say in alphabetical order, with the one-letter strings first, then the two-letter strings, and so on.

2. Separate each string into lines by writing down the segments of it that fall between successive occurrences of the line-separating symbols.

3. By comparing each line with the preceding lines, decide, for each string, whether it is a proof.

4. On a second list, write down each last line of each proof, provided it has the form "The answer to Q_i is A_i."

Thus any sound formal system F is *incomplete* with respect to sentences of the form "The answer to Q_i is A_i": there are some true sentences of this form that F does not prove.

It is true that there are several matters arising from this argument. What is the significance of soundness, or consistency? Can we find specific unprovable sentences? Are there unprovable sentences in mainstream mathematics? All of these questions repay further attention. But for Post the most important message was that incompleteness is an immediate consequence of unsolvability. He also saw that unsolvability is a simple consequence of the diagonal argument (see the subsection below on computable diagonalization).

FORMAL SYSTEMS ARE NORMAL SYSTEMS

The really *big* problem, in Post's view, was to show that all computation is reflected in normal systems. Without a precise definition of computation, the concept of unsolvable problem is meaningless, and there is no general concept of formal system. Post realized that his argument for incompleteness was not really mathematics until computation was defined. At the same time he foresaw that the definition of computability would become a kind of scientific hypothesis, requiring eternal testing and confirmation. For this reason, he did not publish his results, and was scooped by Kurt Gödel, who published a less general, but rigorously proved, incompleteness theorem in 1931. Ever since, the result has been known as *Gödel's incompleteness theorem*.

It could be said that Post travelled further than Gödel along the road to a general incompleteness theorem, but, in traveling so far, he reached an impasse. In the 1920s, logicians were not ready for a general definition of formal system.

Gödel's proof was a specialization of the diagonal argument to the *Principia Mathematica* system. He believed that incompleteness would be provable for "related systems," but he avoided trying to define formal systems in general. Also, he did not see incompleteness as a consequence of unsolvability. He thought that the "true reason" for incompleteness was the ability to form objects of *higher type* than the natural numbers, such as the set of natural numbers. The idea of higher types (or higher *ranks* as we now call them) turns out to be a powerful ingredient for cooking up incompleteness proofs, as we will see in Chapter 7. However, Gödel's 1931 incompleteness proof now looks less like an application of higher types and more like a precursor of the first published unsolvability proofs, by Church in 1935 and Turing in 1936 (see Sections 3.5 to 3.7).

In 1931, the concept of unsolvability was invisible to readers of Gödel's proof, because the concept of computability was generally unknown, and it remained so until 1936. Gödel did not himself *believe* that computability (and hence unsolvability) could be precisely defined until he read the 1936 paper in which the Turing machine concept was defined (see Section 3.5). But after some time it became clear that many approaches to computation, including Post normal systems, define exactly the same concept—which is the concept realized by the electronic computers of today. [3]

COMPUTABLE DIAGONALIZATION

Post defined a set D of positive integers with unsolvable *membership problem*,

$$\langle \text{"Is 1 in } D?\text{"}, \text{"Is 2 in } D?\text{"}, \text{"Is 3 in } D?\text{"}, \ldots \rangle,$$

by a variation of Cantor's diagonal argument. He observed that if S is a set of positive integers with solvable membership problem then one can *compute a list* of the members of S, so S is what we now call *computably enumerable* (c.e.). Assuming that all computation can be realized by normal systems, we can list all c.e. sets $S_1, S_2, S_3, \ldots$. To be specific, S_n can be associated with the nth normal system by letting m be in S_n if the string of m 1s is a "theorem" of the nth normal system (in the listing by size and alphabetical order, say).

The diagonal argument tells us that the set

$$D = \{n : n \text{ is not in } S_n\}$$

[3]In retrospect, we can say that Post's version of *Principia Mathematica* was the first universal computer, though no one knew it at the time. In fact, those were the days in which "computers" were *people* who did computations.

differs from each set S_n, hence it is *not* c.e., and hence its membership problem is unsolvable. It follows, as we saw at the beginning of this section, that no sound formal system can prove all true sentences of the form "n is in D."

This is an easy version of the incompleteness theorem, foreseen by Post in 1921, but first proved rigorously by Gödel in 1930–1931. (As mentioned above, Gödel did not attempt to define all computably enumerable sets; nevertheless his argument is technically rather similar, concerning membership statements provable in *Principia Mathematica*.)

Post also gave a stronger version: *for any sound formal system F, we can find a specific true sentence "$n(F)$ is not in $S_{n(F)}$" not proved by F.* Consider the set

$$D(F) = \{n : F \text{ proves "}n \text{ is not in } S_n\text{"}\},$$

This is a c.e. set, an index $n(F)$ of which we can find by writing down a specific normal system that harvests the numbers n from the theorems of the form "n is not in S_n" produced by the formal system F. Thus

$$S_{n(F)} = \{n : F \text{ proves "}n \text{ is not in } S_n\text{"}\}.$$

Why is it true that $n(F)$ is not in $S_{n(F)}$? Well, if we assume the contrary we get a contradiction as follows:

$$n(F) \text{ in } S_{n(F)} \Rightarrow F \text{ does not prove "}n(F) \text{ is not in } S_{n(F)}\text{"} \text{ because } F \text{ is sound}$$
$$\Rightarrow n(F) \text{ is not in } S_{n(F)} \text{ by definition of } S_{n(F)}.$$

Thus *it is true that $n(F)$ is not in $S_{n(F)}$*. But then it follows, by definition of $S_{n(F)}$, that $n(F)$ is not one of the numbers n for which F proves n is not in $S_{n(F)}$. That is, *F does not prove that $n(F)$ is not in $S_{n(F)}$*.

Gödel also proved incompleteness in this more explicit form, by finding a specific true sentence that the *Principia Mathematica* system does not prove. Moreover, Gödel's approach gives an unprovable sentence in a classical area of mathematics.

3.3 GÖDEL'S FIRST INCOMPLETENESS THEOREM

> I see the essential point of my result not in that one can somehow go outside any formal system (that follows already according to the diagonal procedure), but that for every formal system of mathematics there are statements which are *expressible* within the system but which may *not be decided* from the axioms of that system, and that those statements are even of a relatively simple kind, namely, belonging to the theory of the positive whole numbers.
> —Kurt Gödel,
> Letter to Zermelo, 12 October 1931
> in Gödel (2003), p. 429.

Post's examples of unsolvability and incompleteness are in the theory of formal systems, a field that can express (if not prove) virtually all mathematical statements, though not in the language in which we usually see them. The great contribution of Gödel in 1930–1931 was to bring incompleteness down to earth by showing its existence in *arithmetic*—the theory of positive integers under addition and multiplication. He did this by "arithmetizing" the theory of formal systems, *Principia Mathematica* in particular.

The details of arithmetization are basically straightforward but tedious, so I will only sketch them here.[4] In any case, it is better to see the big picture of arithmetization before zooming in on the details.

First, it is clear that a string of symbols in a formal system F may be interpreted as a positive integer. This is because F has only finitely many symbols, when symbols are realistically counted. For example, a "symbol" such as x_{147} is really a string of several symbols when one takes into account the keystrokes needed to produce it with a computer keyboard. If F has k symbols we can interpret them as the nonzero digits in base $k + 1$ notation. Then each string of symbols of F is a base $k + 1$ numeral that represents a unique positive integer. (We avoid using the digit 0 at this stage, not only to ensure that there are no initial zeros—which would give rise to different numerals representing the same number—but also to save 0 as a *separator* when we come to consider sequences of strings, such as proofs.)

Next, consider the rules of a formal system, which involve recognizing, moving, and changing parts of a string of symbols. Now that we can interpret each string as a number, operations on strings can presumably be interpreted as operations on numbers. But it is not obvious how to describe the typical operations on a *numeral* (such as removing or attaching digits) in terms of the usual operations on the *number* it denotes (such as sum or product). This point needs to be emphasized, because we normally interpret numerals in two ways—sometimes as strings of symbols, sometimes as the numbers these strings denote—without noticing that we are doing two different things. For example, when we say "121 is a three-digit number" we interpret 121 as a string; when we say "121 is a square" we interpret 121 as a number.

What we are about to do is *restore the distinction between the world of numerals and the world of numbers, and establish a precise correspondence*

---

[4]There seems to be no completely natural way to carry out arithmetization. The simplest I know of, which uses only the ideas of numerals and their concatenation, is Smullyan (1961). My sketch of arithmetization uses similar ideas. In Section 7.1 we will discuss a *finite set theory* that is essentially the same as arithmetic. Set theory offers an easier way to "arithmetize" formal systems, though it probably seems easy only to set theorists.

between the two. In particular, we will find operations on numbers that correspond to (or "induce," as we will sometimes say) the operations on numerals that occur in formal systems.

A really strict treatment of numerals would use a separate notation, say $\bar{n}$, for the number denoted by numeral n, (so 121 has three digits, while $\overline{121}$ is a square). But we will trust the reader's experience in distinguishing between numeral and number, and hope that it is clear from context which interpretation is intended. For the sake of familiarity we assume that numerals are in base 10. (However, base b is exactly the same, except that 10 is replaced by b.)

NUMBER FUNCTIONS AND NUMERAL FUNCTIONS

In the world of numbers, the basic functions and relations we will use are the following.

• functions: sum, product, exponential, least number operator,

• relations: less than, equals.

The sum and product functions $m + n$ and mn of numbers are *not* easily described as functions of the numerals for m and n. As you will remember from elementary school, it takes some years to learn how to compute the numerals for $m + n$ and mn from the numerals for m and n. Surprisingly, it is easier to go in the opposite direction!

It is quite simple to find number functions that induce natural operations on numerals. The simplest example is the function $\text{length}(n)$, which counts the number of digits in the numeral for n. Under our assumption that all numerals are written in base 10, we have

$$\text{length}(n) = \text{least } m \text{ such that } 10^m > n.$$

Another important example is the *concatenation function* $m \frown n$ of m and n, which attaches the digits of n to the right of those of m. This function of m and n is

$$m \frown n = m \cdot 10^{\text{length}(n)} + n.$$

For example, $27 \frown 431 = 27 \cdot 10^3 + 431 = 27431$. The concatenation function $\frown$ enables us to induce many operations on numerals by arithmetic operations on numbers.

ARITHMETIZATION OF FORMAL SYSTEMS

The kind of operations we need are those where the numerals are strings in a Post normal system, where production rules are of the type

$$gP \quad \text{produces} \quad Pg'.$$

This is a *numeral* operation, because it assumes that gP and Pg' are strings of symbols. For each production rule "gP produces Pg'" we have a *number* operation that induces it. This is the operation "m produces n" defined by

$$m = g \frown P \text{ and } n = P \frown g' \text{ for some } P.$$

In general, a Post normal system will have finitely many production rules, call them $\text{produces}_1, \text{produces}_2, \ldots, \text{produces}_k$. A numeral n is an *immediate consequence* of numeral m if m produces n by one of the production rules. This numeral relation is induced by the number relation

$$m \text{ produces}_1 n \quad \text{or} \quad m \text{ produces}_2 n \quad \text{or} \quad \ldots \quad \text{or} \quad m \text{ produces}_k n,$$

which we also call immediate consequence (for numbers).

This brings us to the concept of proof. In a Post normal system, a proof is a finite sequence of strings, $s_1, s_2, \ldots, s_l$. The first string s_1 is the axiom, and every other string s_{i+1} is an immediate consequence of s_i. All our strings are zero-free numerals, so we can recover the sequence from the single numeral

$$s_1 0 s_2 0 \ldots 0 s_l.$$

Now, what arithmetic properties of the corresponding number

$$s_1 \frown 0 \frown s_2 \frown 0 \frown \ldots \frown 0 \frown s_l$$

identify it as a proof? To describe these properties we define a property of number m which says that the numeral for m contains no zeros. This property, $\text{NoZerosIn}(m)$, is defined by

$$m \neq p \frown 0 \frown q \quad \text{and} \quad m \neq p \frown 0 \quad \text{for any } p \text{ and } q.$$

The property $\text{NoZerosIn}(m)$ allows us to extract the segment s_1 before the first zero in the numeral for any positive integer n, and also the segments s_i and s_{i+1} between successive zeros. Namely, the segment before the first zero is defined by

$$s_1 = (\text{least } p \text{ such that } n = p \frown 0 \frown q \text{ for some } q),$$

and s_i and s_{i+1} are successive segments between zeros if and only if

$$n = p \frown 0 \frown s_i \frown 0 \frown s_{i+1} \frown 0 \frown q \quad \text{for some } p \text{ and } q$$
$$\text{and} \quad \text{NoZerosIn}(s_i) \quad \text{and} \quad \text{NoZerosIn}(s_{i+1}).$$

Thus the numeral for n is a proof in the formal system F if the initial segment s_1 of n equals the axiom A of F and, for any two successive segments s_i and s_{i+1} of n, s_{i+1} is an immediate consequence of s_i in F.

This property of n is arithmetically definable by what we have established above.

Finally, the numeral for q is a *theorem* of F if and only if the numeral for q is the last line of a proof, that is, the last segment of a numeral that is a proof. The last segment q of a number n is of course defined by

$$\text{least } q \text{ such that } n = p \,^\frown 0 \,^\frown q,$$

so the numeral for q is a theorem of F if and only if there is a positive integer n such that the numeral for n is a proof in F and q is the last segment of n.

Post's Argument Revisited

The results above show that, in a formal system G with a small amount of arithmetic and logic, it is possible to build a "working model" of *any* formal system F. (We call such a system G in honor of Gödel; later we will settle on a more specific system.) In particular, G can express each sentence of the form "m is in S_n" in the language of arithmetic (and prove the sentence if it is true). It follows, by the diagonal argument used in Section 3.2, that *if G is sound then there is a true arithmetical sentence that G does not prove*—namely, the arithmetical equivalent of "$n(G)$ is not in $S_{n(G)}$." This is one form of Gödel's first incompleteness theorem.

The assumption that G is sound is natural, but it turns out to be inconvenient because *we cannot express the soundness of G in the language of G itself*. If we could, we could enumerate all the sets definable in the language of G as $T_1, T_2, T_3, \ldots$ and the set

$$E = \{n : n \text{ is not in } T_n\}$$

would also be definable in G, as, say, T_k. This leads to the contradictions

$$k \text{ in } T_k \Rightarrow k \text{ not in } T_k \quad \text{and} \quad k \text{ not in } T_k \Rightarrow k \text{ in } T_k.$$

This is called the *undefinability of truth*, a remarkable result in itself. It was first published by the Polish (later American) mathematician Alfred Tarski in 1934, though it is mentioned in the 1931 letter from Gödel to Zermelo quoted above.

Looking back, we see that the proof of Gödel's first theorem narrowly avoids contradiction by using the *definable* concept of provability. The gap between provability and truth enables us to *prove* that there is a gap.

Gödel also noticed that the presence of true but unprovable sentences in G could be derived from a weaker assumption than the soundness of G—what he called the *ω-consistency* of G. For the G we have in mind it is enough to assume *simple consistency*, which means that there is no sentence σ such that G proves both σ and $\neg\sigma$ (as mentioned in Section 3.1, $\neg$ stands for "not").

If we review the argument in Section 3.2 that shows "$n(G)$ is not in $S_{n(G)}$" is not provable, we see that it goes through assuming only that G is consistent. We start, as before, by assuming that $n(G)$ is in $S_{n(G)}$, where

$$S_{n(G)} = \{n : G \text{ proves "}n \text{ is not in } S_n\text{"}\}.$$

This assumption means, by the definition of $S_{n(G)}$, that G proves the false sentence "$n(G)$ is not in $S_{n(G)}$". On the other hand, since $n(G)$ is in the c.e. set $S_{n(G)}$, G also proves the sentence "$n(G)$ is in $S_{n(G)}$", thanks to its ability to prove all true membership sentences for c.e. sets. This contradicts the consistency of G, so in fact $n(G)$ is *not* in $S_{n(G)}$. It follows, by the definition of $S_{n(G)}$ again, that G does not prove the sentence "$n(G)$ is not in $S_{n(G)}$".

Thus, *for any formal system G containing a certain amount of arithmetic and logic, the consistency of G implies that the sentence "$n(G)$ is not in $S_{n(G)}$" is true but not provable.* We will not specify what the "certain amount" of arithmetic and logic is, except to say that both addition and multiplication must be present.

3.4 Gödel's Second Incompleteness Theorem

> I have recently concerned myself again with logic, using the methods you have employed so successfully in order to exhibit undecidable propositions. In doing so I have achieved a result that seems to me remarkable. Namely, I was able to show that the consistency of mathematics is unprovable.
> —John von Neumann,
> Letter to Gödel, 20 November 1930,
> in Gödel (2003), p. 337.

Gödel announced his first incompleteness theorem on August 26 1930, at a conference in Königsberg (then in Germany, now the city of Kaliningrad in an isolated part of Russia). Probably the first person to realize its importance was John von Neumann, who wrote to Gödel on November 20, 1930, to point out an extraordinary consequence of Gödel's proof: *the consistency of G is not provable in G.* As it happens, Gödel had already noticed this consequence himself, so it is now known as Gödel's second incompleteness theorem.

Both von Neumann and Gödel had come up with the following argument, which takes the "self-consciousness" of his first incompleteness proof a step further. (Incidentally, like Post's proof of the first incompleteness theorem, Gödel's proof of the second incompleteness theorem was given only as an intuitive argument about manipulations with symbols.)

1. The statement that G is consistent can be expressed as a sentence $\text{Con}(G)$ in the language of G. This is because the set P of provable sentences of G is c.e., G can express membership of any c.e. set, and provability of sentences σ and $\neg \sigma$ can be expressed as membership in P by two numbers with a certain arithmetic relationship.

2. The sentence "G does not prove the sentence '$n(G)$ is not in $S_{n(G)}$' " can also be expressed in G, and it is in fact *equivalent* to the sentence "$n(G)$ is not in $S_{n(G)}$." This follows from the definition

$$S_{n(G)} = \{n : G \text{ proves "} n \text{ is not in } S_n\text{"}\}.$$

(It is commonly said that the sentence "$n(G)$ is not in $S_{n(G)}$" *expresses its own unprovability*.)

3. Therefore, the statement that $\text{Con}(G)$ implies that G does not prove "$n(G)$ is not in $S_{n(G)}$" is equivalent to

$$\text{Con}(G) \Rightarrow \text{"} n(G) \text{ is not in } S_{n(G)}\text{"}$$

This implication is not only expressible in G but *provable* in G, since the argument from consistency of G to unprovability in G is an argument about what happens in formal systems, all of which is expressible in G.

4. But then, if $\text{Con}(G)$ is provable in G, the sentence "$n(G)$ is not in $S_{n(G)}$" will follow, by the modus ponens rule that Q follows from P and $P \Rightarrow Q$. We have established that "$n(G)$ is not in $S_{n(G)}$" is *not* provable in G, so neither is $\text{Con}(G)$.

Unlike the sentence "$n(G)$ is not in $S_{n(G)}$," which is interesting only because it is unprovable, $\text{Con}(G)$ is of great mathematical and logical interest. Thirty years before Gödel, the great German mathematician David Hilbert had posed the problem of proving the consistency of formal systems for mathematics, and arithmetic in particular. His concern was prompted by the paradoxes in set theory, such as the "set of all sets." Hilbert proposed to eliminate the danger of further paradoxes by establishing a formal system for mathematics and proving its consistency by methods he called "finitary."

In the early 1920s, Hilbert and his students (von Neumann among them) made some progress towards this goal, obtaining finitary consistency proofs for some parts of arithmetic. They did not realize that the Achilles heel of full arithmetic was its ability to model *all* formal systems—dooming it to incompleteness and inability to prove its own consistency. Gödel's second incompleteness theorem was virtually the end of Hilbert's program, but it did not end the search for consistency proofs. Rather, it led to an understanding of the relative *strength* or *complexity* of certain sentences of mathematics.

If G is a standard system for arithmetic (which we describe in detail in Chapter 5), then Con(G) is "stronger" or "more complex" than any theorem of G. In 1936 it was discovered by an assistant of Hilbert, Gerhard Gentzen, that the complexity of Con(G) is measured by none other than the ordinal ε_0 that we studied in Chapter 2. Standard arithmetic has an *induction axiom* saying essentially that induction up to the ordinal ω is valid, and it can be proved that this "ω-induction" axiom implies "α-induction" for any $\alpha < \varepsilon_0$. Gentzen proved that Con(G) can be proved by ε_0-induction, thus showing that ε_0-induction is *not* provable in G, and hence that ε_0 is a precise measure of the strength of G. It follows, also, that any sentence that implies ε_0-induction is not provable in G.

One such sentence is a generalization of the Goodstein theorem discussed in Section 2.7. We explain why the generalized Goodstein theorem implies ε_0-induction in Chapter 6. The method of Section 2.7 shows that ε_0-induction also implies the generalized Goodstein theorem, so this theorem is *equivalent* to ε_0-induction, which implies Con(G). Thus, the generalized Goodstein theorem is a true sentence not provable in standard systems of arithmetic, such as G. In fact, the ordinary Goodstein theorem is also not provable in standard arithmetic. The latter theorem dates back to 1944, when ε_0 was known to measure the complexity of arithmetic. However, it was not known that Goodstein's theorem is not provable in standard arithmetic until the late 1970s, when "natural" examples of sentences unprovable in arithmetic were first discovered. All of them were found by taking the "second road to infinity," via ordinals, rather than by the diagonal argument.

3.5 FORMALIZATION OF COMPUTABILITY

Thus the notion 'computable' is in a certain sense 'absolute,' while almost all metamathematical notions otherwise known (for example, provable, definable, and so on) quite essentially depend upon the system adopted.
—Gödel (1936) in Gödel (1986), p. 399.

So far in this chapter we have used normal systems as our model of computation, partly for historical reasons and partly because they exhibit the basic features of formal systems (axioms, rules, proofs, theorems) in a very simple way. However, when it comes to other aspects of *computation*—such as computing real numbers or functions—the easiest model to understand is the Turing machine concept. Turing machines are also of great historical importance, so we include a brief discussion of them in this section and the next.

Gödel proved his first incompleteness theorem for *"Principia Mathematica* and related systems," because a general definition of formal system was not yet available. Post had not published his normal systems, partly because he was unsure that they completely captured the notion of computation. And Gödel, having undermined the idea that *proof* was absolute, doubted that there could be an absolute definition of computation. The first attempt, by the American logician Alonzo Church in 1933, did not convince him.[5]

He was finally convinced by the very intuitive concept of computing machine introduced by the English mathematician Alan Turing in 1936, and now called a Turing machine. Virtually the same concept was introduced independently by Post, at almost the same time. Post was probably moved to publish, at last, because he had seen Church's definition of computation and realized that a definition of greater "psychological fidelity" (as he called it) was needed. Unfortunately, Post's three-page distillation of the concept of computation was too brief to match the psychological fidelity developed at length in Turing's paper.

Turing Machines

We may compare a man in the process of computing a real number to a machine which is only capable of a finite number of conditions $q_1, q_2, \ldots$ The machine is supplied with a "tape" (the analogue of paper) running through it, and divided into sections (called "squares") each capable of bearing a "symbol."
                                                                —Alan Turing (1936), p. 261.

The concept of computing machine introduced by Turing and Post in 1936 is abstracted from the way that human beings compute. (Indeed, it is worth remembering that the word "computer" originally meant a human being who computes.) Turing arrived at his concept by stripping the process of human computation down to its bare essentials: the eye that scans and recognizes symbols, the hand that writes, and the mental states that direct the actions of eye and hand. Assuming that sufficiently similar symbols or mental states will not be distinguished, Turing concludes that

---

[5]Church's computability concept, called *lambda calculus*, also arose from an analysis of the concept of proof—in Church's case from an attempt to construct a formal system for logic and mathematics based on the concept of function. Lambda calculus has been recognized by the computer science community today as a powerful concept in theoretical computer science. However, it remains a rather unfriendly concept for beginners.

computation requires only a finite number of mental states and a finite alphabet of symbols. Further, the symbols can be placed in individual cells and scanned one at a time, because any finite group of symbols can be "remembered" by a mental state that records the individual symbols and their relative positions. Similarly, scanning movements can be broken down into steps from cell to adjacent cell. Finally, the cells can be arranged one-dimensionally—so that the only eye movements are to left and right—since the cells in any array can be given integer addresses, just as if they lay along a line.

Today, we can see an analogy not only with pencil-and-paper computation but with the operation of computing machines. A machine, henceforth called a *Turing machine*, consists of two parts:

1. An infinite *tape* divided into squares, each of which can carry one symbol from a finite alphabet $\{S_0, S_1, S_2, \ldots, S_n\}$. The symbol S_0 is the blank square, and at the beginning of a computation all but finitely many squares are blank.

2. A control mechanism, or *head*, that is in one of finitely many *internal states* $q_1, q_2, \ldots, q_m$. The head scans one square of tape at a time and executes a *step of computation* that depends on the current internal state and the scanned symbol. A step of computation consists of

   - replacement of the scanned symbol by another symbol,
   - move to a neighboring square (left or right),
   - change to a new state.

Thus we can describe a Turing machine by a list of *quintuples* of the form

$$q_i S_j S_k D q_l,$$

where

$$q_i = \text{current state}$$
$$S_j = \text{scanned symbol}$$
$$S_k = \text{replacement symbol}$$
$$D = \text{direction of movement (L or R)}$$
$$q_l = \text{next state.}$$

It is not necessary to have a quintuple for each pair (q_i, S_j). If the machine M reaches a state q_i and symbol S_j for which no action is specified, then M halts.

The tape abstracts the paper on which computations were traditionally written. It is made one-dimensional so as to make the concept of machine as simple as possible—it may be easier to compute in two dimensions but there is no gain in generality. (Think how you would evaluate $1277 + 4123 + 5239$ if forced to work on one line.)

The internal states abstract the mental states of the person doing the computation. In Post's version, internal states are replaced by *instruction numbers* and the quintuple

$$q_i S_j S_k D q_l$$

is replaced by the *instruction*

    i. If the scanned symbol is S_j, replace it by S_k,

        move one square in direction D,

        and go to instruction l.

Thus in Post's formulation, a Turing machine is essentially a *program* in a language that directs the flow entirely by "go to" commands. This is not the most readable programming language, but it reduces computation to remarkably simple ingredients. One can see why Post once thought he was on the brink of solving all mathematical problems.

As we now know, the reality is not that all problems can be solved, but that simply-described problems can be unsolvable. In particular, it is generally not possible to know in advance what ultimately happens when a given program is run.

3.6 THE HALTING PROBLEM

> Before anyone crosses this bridge, he must state on oath where he is going and for what purpose. If he swears truly, he may be allowed to pass; but if he tell a lie, he shall suffer death by hanging ... Now it happened that they once put a man on oath, and he swore that he was going to die on the gallows there.
>   —Miguel de Cervantes
>   *Don Quixote*, Part 2, 51. Translated by J. M. Cohen.

Since each Turing machine T has finitely many states and symbols, T can be described by a finite string of symbols, consisting of a list of the quintuples of T. Moreover, the description can be written in a fixed *finite alphabet* $\{q, S, ', R, L\}$ by encoding each q_i as q followed by i primes, and each S_j as S followed by j primes. Let us call this encoding of the quintuples of T the *standard description* $\lceil T \rceil$ of T. We can similarly encode an input tape sequence I by a word $\lceil I \rceil$ on the alphabet $\{S, '\}$.

Given standard descriptions in a fixed finite alphabet, it is possible (but tedious) to build a *universal Turing machine* U. Given a tape carrying the standard description $\lceil T \rceil$ of a Turing machine T, and the encoding

$\lceil I \rceil$ of the input I for T, U proceeds to "simulate" the computation of T on I. Naturally, the simulation is rather slow, because U has to shuttle back and forth between the part of the tape that encodes T (to find its next instruction) and the part that encodes I (to execute the instruction). In his 1936 paper, Turing gave the first description of a universal machine. Today, most people have one on their desk, because the computers we are familiar with can simulate any computation, assuming an indefinite supply of external storage.

At first glance, it seems as though a universal machine U can answer all questions about computation. It can certainly find out what happens at step t of the computation of machine M on input I, simply by simulating M long enough. But can U decide whether M *halts at some stage*, say, on a blank square? Simply watching U simulate the computation of M on input I may not answer this question because *we do not know when to give up watching*. If halting on a blank square does not occur in the first billion steps of computation, this does not mean it will not occur later.

In fact, there is no Turing machine that correctly answers all of the following questions (as M and I vary).

Does Turing machine M, given input I, ultimately halt on a blank square?

To see why, let us standardize the way in which a Turing machine T may be expected to answer a series of questions. Suppose, for the sake of simplicity, that the questions have YES/NO answers. We make the conventions:

1. If the question concerns machine M with input I then T is given a tape with $\lceil M \rceil$ and $\lceil I \rceil$ on it.

2. T answers YES by halting on a non-blank square, and NO by halting on a blank square.

I think it will be agreed that these are reasonable conventions. Try some other conventions if you prefer. (Note that the machine must halt in order to give an answer, so we need one kind of halting to signal YES, and another kind of halting to signal NO.) It will soon become clear that *any* reasonable conventions for inputting questions and outputting answers can be defeated by a simple problem about Turing machines. In effect, *we turn the conventions against the machine that is supposed to use them*.

Consider the problem that consists of the following questions (which may remind you of the diagonal arguments in Chapter 1, and also of the predicament described in the quote from *Don Quixote*).

Q_M : Does Turing machine M, given its own description $\lceil M \rceil$ as input, ultimately halt on a blank square?

The only parameter in these questions is M, so a machine T that answers them can be expected to take input $\lceil M \rceil$ as the question Q_M and give answer YES by halting on a non-blank square, NO by halting on a blank square. But if so, how does T answer the question Q_T?

If T, given input $\lceil T \rceil$, does not ultimately halt on a blank square, then the answer to Q_T is NO. But then T is supposed to say NO by halting on a blank square! And if T ultimately halts on a blank square then the answer to Q_T is YES, and T is supposed to say YES by halting *not* on a blank square. Clearly, T is stymied. It *cannot* correctly answer the question Q_T.

Thus there is no Turing machine that correctly answers all of the questions Q_M. We say that the problem consisting of the questions Q_M (one version of the so-called *halting problem*) is *unsolvable*.

3.7 The Entscheidungsproblem

> *Wir müssen wissen,*
> *wir werden wissen*
> (We must know,
> we shall know).
>
> —David Hilbert,
> Radio address, November 28, 1930.

As we know from Section 3.2, Post foresaw the existence of unsolvable problems in 1921, along with certain consequences such as the incompleteness of formal systems for mathematics. But Post also recognized that these results had to wait for agreement on the definition of computability—and agreement did not solidify until the publication of Turing's 1936 paper. Reinforcements also arrived in the form of *equivalence proofs*, showing that anything computable by a Turing machine was computable in other formalizations of computation (such as those of Post and Church) and vice versa. The discovery that the Turing machine had many equivalents—and the failure to discover any stronger notion—led to the conviction that

computability = Turing machine computability.

This became known as *Church's thesis*, since it was first proposed by Church.

Church's thesis is assumed whenever one claims that a certain problem (of the type consisting of an infinite set of questions) is unsolvable. This is because one always *proves* that the problem is unsolvable by Turing machine (or one of its equivalents).

The first unsolvability proofs were based on "diagonal" or "self-referential" arguments in some formalization of computability. The argument for the halting problem is a good example. They were followed almost immediately by unsolvable problems in arithmetic or logic, since it was already known how to translate problems about symbol manipulation into arithmetic or logic. The technique of arithmetization sketched in Section 3.3, for example, can be used to reduce the halting problem to the problem of deciding whether certain sentences of arithmetic are true. It follows, by the unsolvability of the halting problem, that the problem of deciding truth in arithmetic is also unsolvable. This result was first published by Church in 1936.

Another problem ripe for an unsolvability proof was the so-called *Entscheidungsproblem* (German for "decision problem"), proposed by Hilbert in 1928. The problem is to decide *validity* in a symbolic language called *predicate logic*. Predicate logic is both a language for expressing logic and mathematics, and a system for proving theorems. We discuss how it works in the next chapter, but its bare bones are the following:

- symbols for variables $x, y, z, \ldots$ and constants $a, b, c, \ldots$,

- symbols for relations ("predicates") in any number of variables,

- symbols for functions of any number of variables.

- symbols for "and," "or," "not," "implies," "for all x," and "there is an x."

All the standard mathematical theories have axioms that can be expressed in predicate logic, and a theorem $\mathcal{T}$ *follows* from axioms $\mathcal{A}_1, \mathcal{A}_2, \ldots, \mathcal{A}_k$ just in case the sentence

$$(\mathcal{A}_1 \text{ and } \mathcal{A}_2 \text{ and } \ldots \text{ and } \mathcal{A}_k) \text{ implies } \mathcal{T}$$

is valid in predicate logic.

Thus the Entscheidungsproblem is the mother of all mathematical problems, in the sense that a solution of the Entscheidungsproblem would decide the provability of any sentence in any finite axiom system (which includes systems like ZF set theory that encompass all other standard mathematical theories). Hilbert realized the Entscheidungsproblem was important because of its all-embracing nature—and he also believed that all mathematical problems are solvable. He once said, in a 1930 radio address that can still be heard on the Internet,[6]"We must know, we shall know."

---

[6]Hilbert's address can be read or heard at various web sites. But since web sites sometimes expire, the best way to find it is to enter the words "hilbert," "radio," and "address" in a search engine.

Of course, Hilbert's hopes were dented as soon as unsolvable problems were discovered. In fact, both Church and Turing reduced their unsolvable problems to predicate logic, thus explicitly proving the unsolvability of the Entscheidungsproblem. However, this does not imply that the truth of any *particular* sentence can never be known. Unsolvability means, rather, that the truth of infinitely many sentences cannot be decided in a *mechanical* manner. Sooner or later, new ideas will be needed. Only in this vaguer sense is it possible that "we shall know."

3.8   Historical Background

Symbolic Logic

> It is obvious that if we could find characters or signs suited for expressing all our thoughts as clearly and as exactly as arithmetic expresses numbers or geometry expresses lines, we could do in all matters, insofar as they are subject to reasoning, all that we can do in arithmetic and geometry.
>   —Gottfried Wilhelm Leibniz
>   *Preface to the General Science, 1677.*

Symbolic logic was foreseen by Leibniz in the 17th century, though he never found time to work out the details. He left only some pioneering investigations in combinatorics, and one of the first computers—a machine that could add, subtract, multiply and divide. The subject did not flourish until the 19th century, with the works of Boole and Frege.

George Boole was an Irish mathematician who took up Leibniz's idea of expressing "thoughts" in a symbolic language, like algebra, and determining the truth of these "thoughts" by computation. He developed his ideas in *The Mathematical Analysis of Logic* in 1847 and, in more leisurely fashion, in his *Laws of Thought* of 1854. His work concerns what we now call *propositional logic*, or *Boolean algebra*: the part of logic that most closely resembles ordinary algebra. It has "variables" (which take the values "true" and "false"), and operations Boole called "logical sum" and "logical product" quite similar to ordinary sum and product. In this part of logic, it is indeed possible to compute the value (true or false) of a proposition by algebra.

Gottlob Frege was a German philosopher interested in the logic and concepts of mathematics. Unlike Boole, who sought a "mathematics of logic," as close as possible to known mathematics, Frege sought a "logic of mathematics" broad enough to capture all mathematical reasoning. Frege's logic is now called *predicate logic* because it contains symbols for *predicates* (properties or relations) as well as symbols for individual mathematical objects, along with an array of logical operations that go far

beyond Boole's "sum" and "product." Deciding whether a proposition
in Boolean algebra is always true is a finite problem, because one can sub-
stitute all possible values of the variables, and compute the truth value
of the proposition for each. But deciding the truth value of a formula in
predicate logic is potentially an infinite problem, because a predicate may
have infinitely many instances. (For example, the predicate "n is prime"
is true for infinitely many n and false for infinitely many n.)

Because of its wide scope, predicate logic does not resemble algebra.
In fact, it resembles nothing so much as mathematics itself. One finds
truths of predicate logic as *theorems*, derived from *axioms of logic* by *rules
of inference*. Frege established all this (albeit with an eccentric symbolism
that never caught on) in his 1879 book *Begriffsschrift* (Concept Writing).
He took as axioms certain obvious general propositions, and as rules of
inference certain obvious ways of deriving new propositions from old.
Where Frege's logic differs from mathematics (as normally practised) is
in being completely formal. His rules are purely *rules for the manipulation
of symbols*, and they can be applied without knowing what the symbols
mean. In this sense, Frege's logic also meets Leibniz's goal of "doing
all that we can do in arithmetic," where even a machine can add and
multiply.

Frege, and his successors such as Whitehead and Russell, wished to
make all mathematical reasoning as precise and reliable as elementary
arithmetic. They introduced symbolic languages for mathematics and
explicit rules of inference to eliminate gaps and unconscious assumptions in proofs. As mentioned in Section 3.1, the *Principia Mathematica*
of Whitehead and Russell was a system that aimed to express and rigorously prove all the theorems of mathematics. As we now know, it
failed. However, it should be mentioned that Frege's system achieves a
goal almost as grandiose. It proves all *logically valid* formulas, that is, the
formulas that are true under all interpretations of the predicate symbols.
This fact—the *completeness* of predicate logic—was shown by Gödel in
1929, and in Section 4.6 we explain how it comes about.

Hilbert's Program

> To reach our goal [of proving consistency], we must make the proofs as such
> the object of our investigation; we are thus compelled to a sort of *proof theory*
> which studies operations with the proofs themselves.
>
> —David Hilbert (1922)
> in the translation of Ewald (1996), vol. II, p. 1127

Second on the list of Hilbert's 23 mathematical problems, posed in 1900,
was the problem of the "compatibility of the arithmetical axioms," by
which he meant axioms for *real* numbers. Hilbert did not doubt that

the axioms for *natural* numbers were consistent, but he was concerned about the use of infinite objects in mathematics, such as the set $\mathbb{R}$ of real numbers. As we know from Section 1.8, set theory had shown that certain infinite objects, such as the "set of all sets," are definitely contradictory, so all infinite objects were suspect. With the development of symbolic logic, however, Hilbert saw a way to justify the use of infinite objects by finite means. He hoped that their use could be shown to be free of contradiction by analysis of symbol manipulation.[7]

This train of thought led to the so-called *Hilbert program* for proving the consistency of mathematical theories. It envisioned roughly the following path to glory:

1. Express statements about infinite objects as finite strings of symbols in a precisely-defined formal language.

2. Express rules of inference as precisely-defined operations on symbol strings.

3. Derive theorems from the axiom strings by finitely many applications of rules of inference.

4. Prove that the rules of inference produce no contradictory sentence, by purely *finite* reasoning about strings of symbols.

Of course, steps 1 to 3 were carried out by logicians such as Frege and Russell. Step 4 looked like a job for the theory of natural numbers, since strings of symbols are like numerals and simple functions of natural numbers realize basic operations. What Hilbert overlooked was that the outcome of even simple operations on numerals can be hard to foresee. Hilbert's program, under a reasonable interpretation of finite reasoning, is doomed by Gödel's theorems.

This does not mean, however, that formalization is a waste of time or that consistency proofs are vacuous. The Hilbert program has at least two positive outcomes.

- Formal proofs, like numerical calculations, can be checked by machine. Therefore, if we can generate formal proofs of theorems whose *in*formal proofs are hard to follow, we can guarantee that

---

[7]It is commonly thought that Hilbert was a "formalist" who wished to purge intuition from mathematics in favor of pure symbol manipulation. Actually, Hilbert was a great believer in intuition, but he recognized that symbol manipulation is the area where *everyone has the same intuition*. (Or, at least, everyone who can read—which presumably covers the audience for mathematics!) All proofs should, therefore, be reduced to symbol manipulations if we want everyone to accept our conclusions. In the Hilbert (1922) paper quoted above, p. 1122, Hilbert expressed this view by the slogan: *In the beginning was the sign.*

these theorems are correct. This has recently been done for some famous hard theorems, such as the prime number theorem (by Jeremy
Avigad) and the four-color theorem (by Georges Gonthier). In the
future, we may obtain machine-checkable proofs of theorems whose
proofs are virtually impossible for any one person to grasp, such as
the classification theorem for finite simple groups.

- Consistency proofs, even though they involve infinite reasoning,
  may bring to light *interesting* statements about infinity. This is the
  case with Gentzen's consistency proof for Peano arithmetic, or PA
  (see Chapter 5), which brings to light the principle of ε_0-induction.
  Since ε_0-induction implies Con(PA), it is not a theorem of PA, by
  Gödel's second theorem. This in itself makes ε_0-induction an interesting statement. It is also interesting because of other statements it
  implies, such as the Goodstein theorem of Section 2.7.

Precursors and Successors of Gödel

Post is the most important precursor of Gödel (and Turing) because he
clearly grasped the essential ideas of incompleteness and unsolvability,
and the connection between them. It is true that we only have Post's
word for this, but his later publications confirm his insight and original-
ity. As we know, he published the Turing machine concept independently
of Turing in 1936. He also published a watershed paper on computably
enumerable sets in 1944, and in 1947 discovered the first example of an
unsolvable problem in algebra (the word problem for semigroups, also
discovered independently by A.A. Markov). A plausible explanation for
Post's independent discovery of Gödel incompleteness is that the diagonal argument had been in the air for decades, and it was bound to be
applied in the setting of formal systems sooner or later. Post's works on
incompleteness and unsolvability are collected in the volume edited by
Martin Davis (Post (1994)).

There is evidence that the great topologist (and maverick logician)
L. E. J. Brouwer anticipated the incompleteness of formal systems as early
as 1907. Brouwer saw the diagonal argument as proof, not of uncountability, but rather of the eternally "unfinished" nature of the set of the
real numbers. Carrying this idea further, he wrote in one of his note-
books that "The totality of mathematical theorems is, among other things,
also a set which is denumerable but never finished." However, the lesson Brouwer drew from this is that intuition is the only proper basis for
mathematics, and *nothing* can be completely formalized, including the
concept of computation. (For more on Brouwer's early brush with incompleteness, which may even have influenced Gödel, see van Atten and

Kennedy (2009), pp. 498–499.) As we have seen, Gödel himself was at first reluctant to believe that computation could be formalized, but later he reversed his position.

Another flawed anticipation of Gödel incompleteness was made, and published, by Paul Finsler in 1926. Finsler was not interested in incompleteness relative to particular formal systems, but in *absolutely* unprovable results. However, the lesson of Gödel incompleteness is that there is no absolute notion of proof, so it is difficult to make sense of Finsler's claims. (In fairness to Finsler, it should be mentioned that his paper was cited by the great logician Gerhard Gentzen, whose work we will discuss in Chapter 5. Gentzen seems to have viewed Finsler as the discoverer of incompleteness.)

After Gödel and Turing, research on incompleteness stagnated for some time, but research on unsolvability flourished. The papers of Post in 1944 and 1947 launched a thousand investigations into the fine structure of unsolvable problems and a search for them in all parts of mathematics. Following the 1947 breakthrough into semigroup theory, the Russian school of A.A. Markov, P.S. Novikov, and S.I. Adian showed the unsolvability of some fundamental problems in group theory and topology (such as recognizing whether two groups are isomorphic, or whether two manifolds are homeomorphic). Perhaps the climax of this story was the discovery by Yuri Matiyasevich in 1970 that *Hilbert's tenth problem is unsolvable*. The tenth problem on Hilbert's list in 1900 was to find an algorithm which decides, for any polynomial $p(x_1, x_2, \ldots, x_n)$ with integer coefficients, whether the equation $p(x_1, x_2, \ldots, x_n) = 0$ has an integer solution. Matiyasevich showed that there is no such algorithm, and it follows, by Post's argument, that any consistent formal system for mathematics fails to prove some theorem of the form

$$\text{"}p(x_1, x_2, \ldots, x_n) = 0 \text{ has no solution."}$$

In the 1960s and 1970s, Gödel incompleteness made a comeback, with the discovery of new types of unprovable sentences in arithmetic.[8] The examples from the 1970s (and later) will be discussed in Chapter 6. Here we mention only an example from the 1960s.

In the 1960s, several mathematicians independently discovered that there is an interesting connection between computation and the "randomness" or "incompressibility" of finite strings of symbols. Today, the idea of "compression" is familiar to many people, since computer files

---

[8]In a separate development, incompleteness was discovered in set theory. We have mentioned the examples of the axiom of choice and the continuum hypothesis in Chapter 2. We call these sentences "undecidable," rather than "unprovable," because we are unsure what is true—the sentences or their negations. Other sentences of set theory, whose unprovability is quite easy to show, will be discussed in Chapter 7.

are often "compressed" for easier storage. It is also familiar that the more "random" the file, the less it can be compressed—a digital photo of a crowd of people, for example, will yield a larger file than a photo of a blank wall. One expects that a truly random file will not be compressible at all. To take a simple example, a string of 1,000,000 consecutive zeros is surely compressible, because we have just described it (in English) in about 20 symbols, and a Turing machine program that generates it will not be vastly longer (maybe a few hundred symbols). But a random sequence—say, 1,000,000 consecutive coin tosses—will not be describable by any sequence shorter than itself.

The general idea may be formalized as follows. Consider all finite strings of 0s and 1s, and all programs in some universal programming language, such as Turing machine programs. We call a string S of n symbols *incompressible* if it is not the output of a program of length less than n. To make a fair comparison between programs and their output strings, let us assume that the programs themselves are encoded as strings of 0s and 1s. Then it follows that *incompressible strings exist*, simply because

- there are 2^n strings of length n, and

- there are at most $2^{n-1} + 2^{n-2} + \cdots + 2 + 1 = 2^n - 1$ programs of length $< n$.

Thus there are infinitely many incompressible strings—at least one of each length n. Despite this, *no sound formal system can prove more than finitely many theorems of the form "string S is incompressible."* This startling

---

CHAITIN'S INCOMPLETENESS THEOREM

Any sound formal system proves only finitely many theorems of the form "string S is incompressible."

Suppose, on the contrary, that M is a sound formal system that produces infinitely many theorems of the form "string S is incompressible." We may take M to be a Turing machine program, whose encoded form has length, say, 10^6 symbols. Since M finds infinitely many incompressible strings, we can modify it to a machine M' that searches through the strings proved incompressible by M until it finds one, S', of length at least 10^9. Clearly, the encoded form of M' is not much longer than that of M. But then the string S' is generated by the program M' shorter than S', contrary to the incompressibility of S'.

---

form of incompleteness was discovered by Gregory Chaitin in 1970, following his discovery of incompressible strings in 1965 as a student in New York. The proof is surprisingly easy (see previous page).

Chaitin's incompleteness differs from that of Gödel (or Post) in that we cannot uniformly compute an unprovable sentence of the form "string S is incompressible" from the formal system M. If we could, then we could iterate the computation to generate infinitely many such sentences—contradicting incompressibility as in the proof of Chaitin's theorem itself.

Chaitin's argument also yields a curious unsolvable problem: *deciding which strings of 0s and 1s are compressible by Turing machine programs.* This problem has the remarkable feature that we can mechanically find all of the "yes" instances (by running all Turing machine programs and harvesting the outputs that are 0–1 strings longer than the programs themselves), but only finitely many of the "no" instances. Post devised such problems in his 1944 paper, but Chaitin's was the first example for which the property to be decided has independent interest.

A convenient source for most of the basic papers of Gödel, Church, Post and Turing, including the unpublished paper Post (1941) that describes his anticipation of Gödel and Turing, is Davis (2004). A rather different selection, including the flawed paper of Finsler, is in van Heijenoort (1967).

- CHAPTER 4 -

LOGIC

PREVIEW

Gödel's second incompleteness theorem seems to tell us that it is futile to try to prove the consistency of arithmetic. But why not try anyway? Something interesting may happen, and it does. To explain what, we need to give a precise description of formal logic and its extension to formal arithmetic. In this chapter we formalize the languages needed to express logic and arithmetic, and the rules needed to generate proofs in the relevant languages, then reconsider the questions of completeness and consistency.

Historically, the most important rule of logic is *modus ponens*, the ancient rule that if A holds, and A implies B, then we may infer B. The modern version of modus ponens is called *cut*, because the sentence A in the premises is cut out of the conclusion. Despite its pedigree, the cut rule is actually the "wrong" way to do logic, because pure logic can be done more simply without it. We explain how in Sections 4.3 and 4.4.

We separate logic into *propositional logic* (the logic of "and," "or," and "not"), and *predicate logic* (the logic of relations between individuals $x, y, z, \ldots$ and the *quantifiers* "for all x" and "there exists an x"). Propositional logic is in some sense the "finite part" of predicate logic, so it is natural to study it first. Then the key property of predicate logic—its completeness—can be seen as a natural extension of completeness for propositional logic.

Logic is complete in the sense that we have a consistent formal system that proves all sentences that are valid, that is, true under all interpretations. Consistency of logic is relatively trivial, but completeness is quite remarkable (and of course in sharp contrast to the case of arithmetic).

4.1   PROPOSITIONAL LOGIC

The simplest kind of logic is the logic of propositions, or propositional logic. Propositions (also known as sentences) are utterances that have a definite *truth value,* true or false. Thus "$1 + 1 = 3$" is a proposition, but the question "Is $1 + 1 = 3$?" is not, and neither is the command "Let $1 + 1 = 3$." In propositional logic, the internal structure of propositions is not analyzed, so they are denoted by variables such as $P, Q, R, \ldots$ whose only characteristic is that they can take the values "true" and "false." Such two-valued variables are known as *Boolean variables,* after Boole, who introduced this approach to propositional logic in his book *The Mathematical Analysis of Logic* in 1847.

Since each variable can take only two values, a Boolean-valued function f of Boolean variables can be specified by a finite table, giving the value of f for each of the finitely many combinations of values of its variables. Here are the tables for some important Boolean functions: a function of one variable denoted by $\neg P$, and functions of two variables denoted by $P \vee Q$ and $P \wedge Q$. We denote the values "true" and "false" by T and F respectively.

P	$\neg P$
F	T
T	F

P	Q	$P \vee Q$
F	F	F
F	T	T
T	F	T
T	T	T

P	Q	$P \wedge Q$
F	F	F
F	T	F
T	F	F
T	T	T

The function $\neg P$ is of course the negation function "not P" that we introduced in Section 3.1. The function $P \vee Q$ is the function "P or Q," where "or" is understood in the inclusive sense that is usual in mathematics: "P or Q or both." $P \vee Q$ is also called the *disjunction* of P and Q. The function $P \wedge Q$ is the function "P and Q," also called the *conjunction* of P and Q. The tables are called *truth tables.* The functions $\wedge$, $\vee$, and $\neg$ interpret propositional logic in ordinary speech, and we can take advantage of this interpretation to show that *any Boolean function* can be expressed in terms of "and," "or," and "not."

One example should suffice to explain why this is so. Consider the function $f(P, Q)$ given by the following truth table.

P	Q	$f(P, Q)$
F	F	F
F	T	T
T	F	F
T	T	T

This table says that $f(P,Q)$ is true if and only if

(P is not true and Q is true) or (P is true and Q is true).

Thus $f(P,Q)$ is the same as the function $((\neg P) \wedge Q) \vee (P \wedge Q)$. Clearly, any truth table can be similarly expressed in terms of $\wedge$, $\vee$, and $\neg$: just write down the conjunction of variables and negated variables for each line that gives the value T, then take the disjunction of these conjunctions.

A particularly important Boolean function for mathematics is the function $P \Rightarrow Q$, expressed by "If P then Q" or "P implies Q" in ordinary speech. Its truth table is

P	Q	$P \Rightarrow Q$
F	F	T
F	T	T
T	F	F
T	T	T

which happens to be the same as the truth table for $(\neg P) \vee Q$. In fact, *every Boolean function can be expressed in terms of $\neg$ and $\vee$.* We already know that every Boolean function can be expressed in terms of $\wedge$, $\vee$, and $\neg$. Also, $\wedge$ can be expressed in terms of $\neg$ and $\vee$ because "P and Q" holds if and only if "it is false that P is false or Q is false," that is:

$$P \wedge Q = \neg((\neg P) \vee (\neg Q)).$$

Thus all Boolean functions can be expressed in terms of $\neg$ and $\vee$ alone. It is in fact possible to express all Boolean functions in terms of a single function of two variables (for example the "nor" function $(\neg P) \wedge (\neg Q)$), but there are not many benefits in doing so. The pair of functions $\{\neg, \vee\}$ gives a particularly nice formal system for propositional logic, as we will see in the next section.

What is the main problem of propositional logic? An obvious candidate is the problem of deciding whether a given sentence, written in terms of variables $P_1, P_2, \ldots, P_n$ and the functions $\wedge$, $\vee$, and $\neg$ is valid, that is, true under all interpretations of the variables. Since the n variables $P_1, P_2, \ldots, P_n$ can be assigned values T or F in $2 \times 2 \times \cdots 2 = 2^n$ different ways, the validity problem can be solved by computing the value of the sentence (using the truth tables for $\wedge$, $\vee$, and $\neg$) for each assignment of values. For example, here is a computation showing that $P \vee (\neg P)$ has value T for each value of the variable P, so $P \vee (\neg P)$ is valid.

P	$\neg P$	$P \vee (\neg P)$
F	T	T
T	F	T

This method also solves the *satisfiability problem*, which is to decide whether a sentence has the value T for *some* assignment of value to the variables. It is certainly nice to have a finite solution to both of these problems, though it is not typical of logic or mathematics in general, where the process of evaluation is usually infinite.

For an understanding of logic in general it is better to find a formal system for propositional logic, which produces the valid sentences as theorems. We look at such formal systems in the next two sections.

4.2   A Classical System

> The reader ... may be led to remark that this proposed theorem is not only obvious but more obvious than any of the axioms.
>
> —Alonzo Church (1956), p. 81.

As mentioned above, classical formal systems for propositional logic rely heavily on the modus ponens or cut rule,

$$\text{from } P \Rightarrow Q \text{ and } P \text{ infer } Q,$$

to prove theorems. Such systems originated in the 1880s with Frege, and they were popularized in the monumental *Principia Mathematica* of Whitehead and Russell, the first volume of which appeared in 1910.

A typical system of this type appears in the *Introduction to Mathematical Logic* of Church (1956). It has the following three axioms:

$$P \Rightarrow (Q \Rightarrow P) \qquad \text{(affirmation of the consequent)}$$
$$((S \Rightarrow (P \Rightarrow Q)) \Rightarrow ((S \Rightarrow P) \Rightarrow (S \Rightarrow Q)))$$
$$\text{(self-distribution of implication)}$$
$$\neg\neg P \Rightarrow P \qquad \text{(double negation)}$$

As Church suggests in the quotation above, these axioms are not entirely obvious. However, it is possible to check their validity by truth tables. It is then clear that we have validity of their *consequences* by the cut rule and a rule of substitution that allows P, Q, or S to be replaced (everywhere it occurs) by any sentence built from variables and the $\neg$ and $\Rightarrow$ symbols.

So far, so good. But try proving something, say, the "obvious" sentence $P \Rightarrow P$. A proof is not just hard to find—it is hard to read, even if someone finds it for you:

$$(P \Rightarrow (Q \Rightarrow P)) \Rightarrow ((P \Rightarrow Q) \Rightarrow (P \Rightarrow P))$$
$$P \Rightarrow (Q \Rightarrow P)$$
$$(P \Rightarrow Q) \Rightarrow (P \Rightarrow P)$$
$$(P \Rightarrow (Q \Rightarrow P)) \Rightarrow (P \Rightarrow P)$$
$$P \Rightarrow P$$

Here is an explanation (minus some substitutions we leave to the reader).

- Line 1 is a substitution instance of the second axiom.

- Line 2 is the first axiom.

- Line 3 comes from lines 2 and 1 by cut.

- Line 4 is a substitution instance of line 3.

- Line 5 comes from lines 2 and 4 by cut.

Mathematical readers may not be surprised at making a detour through a series of long sentences in order to prove a short one. After all, it is quite common in mathematics for a short sentence to have a long proof. (For example, the sentence "$x^n + y^n \neq z^n$ for positive integers x, y, z and $n > 2$.") This may be why logicians tolerated this formal system for so long (and still do). But *why take a detour when there is a direct route?* We already know a direct method for deciding validity (truth tables), and in fact there is also a direct method for proving valid sentences.

The key to making proofs more direct is to do without the cut rule. As we can see from the proof above, when cuts are allowed the proof of a theorem S may involve long intermediate sentences not obviously related to S. Ideally, the proof of S should move towards S through sentences simpler than S. In particular, *the axioms should be the simplest possible valid sentences*, namely, sentences of the form $P \Rightarrow P$.

There is in fact a cut-free formal system for propositional logic based on the axioms $P \Rightarrow P$, which we write in terms of $\neg$ and $\vee$ as $\neg P \vee P$ (or $P \vee \neg P$, since the order does not matter). We choose $\neg$ and $\vee$ to express all Boolean functions not because it is essential, but because it gives a particularly simple system, as we will see in the next section. The simplicity emerges when one realizes that proofs are *not* naturally linear, but, rather, *tree-shaped*.

(See Figure 4.1, which is Caspar David Friedrich's *Tree of Crows*.) In logic, the tree shape arises from trying all possible ways to *falsify* the sentence S, which turns out to be a better idea than trying to prove S directly.

Figure 4.1. The natural shape of a proof.

4.3 A Cut-Free System for Propositional Logic

> Perhaps we may express the essential properties of a normal proof by saying:
> it is not roundabout.
>
> —Gerhard Gentzen (1935)
> Translation by M.E. Szabo in Gentzen (1969), p. 69.

In Section 4.1 we saw that any valid sentence of propositional logic can be recognized by a process that begins with the sentence and ends with a collection of very simple things—the values T of the sentence under each assignment of values to its variables. It is harder to see how one might begin with some simple things (such as axioms) and end up with an arbitrary valid sentence, as is supposed to happen in a formal system. It appears easier to *work backwards*, from sentence to axioms, so that is what we will do.

Given a sentence S in the language with $\neg$, $\vee$ and variables $P_1, P_2, \ldots$, we try to *falsify* S by finding parts of S whose falseness (for some values to the variables) would make S false. If we fail, by reaching formulas that obviously cannot be falsified—namely, axioms—then we can conclude that S is valid. Then, by reversing the attempted falsification, we obtain a proof of S.

Given that all sentences are Boolean functions built from $\neg$ and $\vee$, any sentence S is either a variable, of the form $\mathcal{A} \vee \mathcal{B}$, or of the form $\neg \mathcal{C}$.

Figure 4.2. Neighboring sentences in a falsification tree.

$\mathcal{A} \vee \mathcal{B}$ can be falsified only by falsifying both $\mathcal{A}$ *and* $\mathcal{B}$, in which case we keep these two parts together and look inside them. Likewise, to decide whether $\neg\mathcal{C}$ can be falsified, we look inside $\mathcal{C}$ and act as follows, according as $\mathcal{C}$ is $\neg\mathcal{A}$ or $\mathcal{A}_1 \vee \mathcal{A}_2$.

1. A sentence of the form $\neg\neg\mathcal{A}$ is falsified by falsifying $\mathcal{A}$.
   (Because if $\mathcal{A} = \mathsf{F}$ then $\neg\neg\mathcal{A} = \neg\neg\mathsf{F} = \mathsf{F}$.)

2. A sentence of the form $\neg(\mathcal{A}_1 \vee \mathcal{A}_2)$ is falsified by falsifying $\neg\mathcal{A}_1$ or $\neg\mathcal{A}_2$.
   (Because if, say, $\neg\mathcal{A}_1 = \mathsf{F}$ then $\neg(\mathcal{A}_1 \vee \mathcal{A}_2) = \neg(\mathsf{T} \vee \mathcal{A}_2) = \neg\mathsf{T} = \mathsf{F}$.)

More generally, we can carry along the disjunction with any sentence $\mathcal{B}$, and state falsifiability in both directions, namely:

$\neg\neg$-**falsification.** An assignment makes $\neg\neg\mathcal{A} \vee \mathcal{B}$ false if and only if it makes $\mathcal{A} \vee \mathcal{B}$ false.

$\neg\vee$-**falsification.** An assignment makes $\neg(\mathcal{A}_1 \vee \mathcal{A}_2) \vee \mathcal{B}$ false if and only if it makes at least one of $\neg\mathcal{A}_1 \vee \mathcal{B}$ or $\neg\mathcal{A}_2 \vee \mathcal{B}$ false.

We apply the rules diagrammatically as shown in Figure 4.2: the sentence in the top line is falsifiable if and only if one of the sentences below it is falsifiable. Applying these rules to any sentence $\mathcal{S}$ gives a *tree* with root $\mathcal{S}$ and downward branches, ending in sentences with no occurrence of $\neg\neg\mathcal{A}$ or $\neg(\mathcal{A}_1 \vee \mathcal{A}_2)$. This means that the sentence at the end of each branch is simply a disjunction of variables or negated variables, and we can falsify $\mathcal{S}$ if and only if we can falsify some sentence at the end of a branch.

EXAMPLE. $\mathcal{S} = (P \Rightarrow Q) \Rightarrow (\neg Q \Rightarrow \neg P)$
   Rewriting each $A \Rightarrow B$ as $(\neg A) \vee B$, and omitting parentheses where possible, we obtain

$$\mathcal{S} = \neg(\neg P \vee Q) \vee (\neg\neg Q \vee \neg P) = \neg(\neg P \vee Q) \vee \neg\neg Q \vee \neg P.$$

Then, applying $\neg\neg$- and $\neg\vee$-falsification to the latter sentence, we obtain the tree shown in Figure 4.3, with root $\mathcal{S} = \neg(\neg P \vee Q) \vee \neg\neg Q \vee \neg P$. The

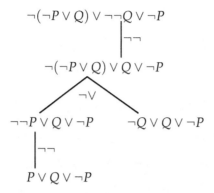

Figure 4.3. Example of a falsification tree.

sentences at the ends of the branches are $P \lor Q \lor \neg P$ and $\neg Q \lor Q \lor \neg P$. Neither of these can be falsified, because each contains a variable in both forms—with and without a negation sign. Thus the original sentence $\mathcal{S}$ is valid.

Clearly, a disjunction of variables and negated variables can be falsified *only* if each variable occurs in only one form. For example, $P \lor \neg Q \lor P$ can be falsified by assigning the values $P = \mathsf{F}$, $Q = \mathsf{T}$. Thus *a sentence $\mathcal{S}$ is valid if and only if each branch produced by $\neg\neg$- and $\neg\lor$-falsification ends in a sentence in which some variable appears both with and without a negation sign.*

Now consider the reverse of this process. We take as:

AXIOMS. All disjunctions of variables and negated variables in which some variable appears both with and without a negation sign.

RULES. Directed versions of $\neg\neg$- and $\neg\lor$-falsification, namely:

$\neg\neg$**-rule.** From $\mathcal{A} \lor \mathcal{B}$ infer $\neg\neg\mathcal{A} \lor \mathcal{B}$.

$\neg\lor$**-rule.** From $\neg\mathcal{A}_1 \lor \mathcal{B}$ and $\neg\mathcal{A}_2 \lor \mathcal{B}$ infer $\neg(\mathcal{A}_1 \lor \mathcal{A}_2) \lor \mathcal{B}$.

If we apply $\neg\neg$- and $\neg\lor$-falsification to a valid sentence $\mathcal{S}$ so as to remove all occurrences of $\neg\neg\mathcal{A}$ and $\neg(\mathcal{A}_1 \lor \mathcal{A}_2)$, the result is a tree whose branches end in axioms. Therefore, if we read the falsification tree from bottom to top, the result is a *proof* of $\mathcal{S}$ by the $\neg\neg$- and $\neg\lor$-rules. Our axioms and rules are therefore *complete* for propositional logic: they prove all valid sentences built from the functions $\neg$ and $\lor$, which, as we have seen, suffice to express all Boolean functions.

It must be admitted that this system of axioms and rules is not completely explicit. We have quietly assumed that it is permissible to ignore parentheses and the order of terms in a disjunction. These assumptions can be formalized by extra rules:

Regrouping. From $\mathcal{A} \lor (\mathcal{B} \lor \mathcal{C})$ infer $(\mathcal{A} \lor \mathcal{B}) \lor \mathcal{C}$, and vice versa.

Exchange. From $\mathcal{A} \lor \mathcal{B}$ infer $\mathcal{B} \lor \mathcal{A}$.

Also, we can simplify the axioms to be all sentences of the form $P_i \lor \neg P_i$ if we include the so-called *dilution* rule:

Dilution. From $\mathcal{A}$ infer $\mathcal{A} \lor \mathcal{B}$.

Finally, we can ignore repetitions by the *consolidation* rule:

Consolidation. From $\mathcal{A} \lor \mathcal{A} \lor \mathcal{B}$ infer $\mathcal{A} \lor \mathcal{B}$.

These rules of inference have the desirable property that the premises are in some sense "contained in" the conclusion. Unlike proofs using the cut rule, there are no "detours" through sentences more complicated than the sentence we are trying to prove.

4.4 Happy Endings

Generating proofs simply by running the falsification procedure backwards enables us to tailor proofs to theorems to some extent. In particular, we can guarantee that the last step of a proof is a "happy ending" that yields the target theorem in the most natural way. Namely,

1. If the target theorem is of the form $\neg\neg\mathcal{A} \lor \mathcal{B}$, then the last step is

   from $\mathcal{A} \lor \mathcal{B}$ infer $\neg\neg\mathcal{A} \lor \mathcal{B}$.

2. If the target theorem is of the form $\neg(\mathcal{A}_1 \lor \mathcal{A}_2) \lor \mathcal{B}$, then the last step is

   from $\neg\mathcal{A}_1 \lor \mathcal{B}$ and $\neg\mathcal{A}_2 \lor \mathcal{B}$ infer $\neg(\mathcal{A}_1 \lor \mathcal{A}_2) \lor \mathcal{B}$.

We can guarantee the last step in the proof of a theorem $\mathcal{S}$ by choosing its reverse as the *first* step in an attempt to falsify $\mathcal{S}$. Since $\mathcal{S}$ is a theorem, *any* attempt to falsify it yields a tree whose branches end in unfalsifiable formulas. Reading this tree backwards gives a proof of $\mathcal{S}$ with the desired ending.

This property becomes important in systems with extra ingredients, such as the cut rule. In the next chapter we will find circumstances in which cuts may be eliminated by studying the first occurrence of the cut rule and arranging "happy endings" for the cut-free proofs down to that point in the tree.

CONSISTENCY OF PROPOSITIONAL LOGIC

The method of evaluation by truth tables is not entirely superseded by the formal system described by the axioms $P_i \lor \neg P_i$ and the rules above. The formal system gives us all the valid sentences as theorems, but now we need to know that all theorems are valid; that is, the soundness of the system. Evaluation is just what we need to prove that *the axiom system for propositional logic is sound (and hence consistent).*

Proof: First, it is clear that each axiom has the constant value T. This was our example of evaluation in the previous section. Second, it can be checked that, if the premises of a rule have the constant value T, then so has the conclusion. Thus each theorem has the constant value T, and hence the system is sound—it proves only valid sentences.   □

Propositional logic is presumably not expressive enough for the proof of Gödel's second incompleteness theorem to apply to it. Nevertheless, the consistency proof just described *does* depend on a mathematical principle beyond those present in propositional logic itself; namely, induction for finite ordinals.

To prove that a given theorem S is valid we essentially assume that all shorter theorems are valid, in which case S is valid because it is necessarily inferred from shorter premises. (In the case of the regrouping and exchange rules we have to step back to the rule used *before* regrouping or exchange was applied.) Thus the consistency proof depends on the principle that any descending sequence of finite ordinals is finite, that is, induction for finite ordinals.

This situation is roughly analogous to what we will later see for arithmetic. The formal system for arithmetic includes induction for finite ordinals, and to prove its consistency we need a stronger induction principle—induction for ordinals less than ε_0.

4.5 PREDICATE LOGIC

To discuss individual objects, properties, and functions, as we do in arithmetic or set theory, we need a language and logic that includes these things. The simplest logic that suffices for mathematics is called *predicate logic*.[1] It has the usual propositional functions, but it also has names for individuals (both *constants* and *variables*), relations (between any finite

---

[1]"Predicates" are what we nowadays call *properties*, which are a special case of *relations* (the one-variable case). It would be better to use the term "relation logic," but this has never caught on, outside a period in the 19th century when logicians spoke of the "logic of relatives."

number of variables), and functions (also of any finite number of variables). Also, for any variable x we have the quantifiers $\forall x$ "for all x" and $\exists x$ "there exists an x" or simply "there is an x." Predicate logic is also called *first order logic* because it permits only quantification over individuals, not over relations or functions (which would give *second order logic*). This sounds like a severe restriction, but first order logic is adequate for most mathematical purposes, and it has better properties than second order logic.

EXAMPLE: A LANGUAGE FOR ARITHMETIC

If the intended individuals are the positive integers we usually have:

CONSTANTS: $1, 2, 3, 4, \ldots$

VARIABLES: $i, j, k, l, m, n, \ldots$

FUNCTIONS: $S(n)$ (successor, a function of one variable n),
   $m + n$ (sum, a function of two variables m and n),
   $m \cdot n$ (product, a function of two variables m and n).

RELATIONS: $m = n$ (a relation of two variables m and n),
   $m < n$ (a relation of two variables m and n).

In this language we can express all the properties of positive integers usually considered in number theory. For example, "n is prime" is the property

$$(n > 1) \wedge \neg(\exists l)(\exists m)(l > 1 \wedge m > 1 \wedge n = m \cdot l).$$

We cannot call this expression for "n is prime" a sentence, because it is neither true nor false. We call it a *formula* with *free variable n*. A formula with free variables $m, n, \ldots$ becomes a sentence when the variables are *bound* by quantifiers. For example, the formula

$$n < m \wedge \text{"}m\text{ is prime"}$$

is a formula with free variables m and n. When we bind the variables like so—

$$(\forall n)(\exists m)(n < m \wedge \text{"}m\text{ is prime"})$$

—we have a sentence equivalent to "There are infinitely many primes."

   If we really want to interpret a formula with free variables as a sentence, then we do so by taking the formula as an assertion for *all* values of its free variables. This is commonly done in mathematics. For example, when we write $m + n = n + m$ it is understood to mean "for all m, for all n."

A Formal System for Predicate Logic

To be prepared for all eventualities, the language for predicate logic should contain infinitely many constants (symbols to denote specific objects, as yet undetermined), infinitely many variables (symbols x for which we can say "for all x" or "there is an x"), infinitely many function symbols, infinitely many relation symbols, and the equality sign. However, we can manage with less than this, by including equality and functions among the relations, and we will do so for the rest of this section.

A surprising fact, discovered by Gödel shortly before he discovered the incompleteness of arithmetic, is that predicate logic is *complete*. That is, there is a formal system whose theorems are precisely the valid sentences of predicate logic. As in propositional logic, "valid" means "true under all interpretations." There are some sentences of predicate logic that are valid for propositional reasons—for example $P(a) \vee \neg P(a)$—but others involve variables and quantifiers in an essential way, for example $(\forall x)P(x) \vee (\exists x)\neg P(x)$. The latter sentence is valid because it is true for all domains the variable x may range over, and for all interpretations of the property P on that domain.

We find a complete formal system for predicate logic by the same strategy as for propositional logic: seeking to falsify sentences by falsifying simpler sentences, then writing the falsification rules backward as rules of inference. As before, we suppose that all Boolean functions have been written in terms of $\neg$ and $\vee$, so the reverse $\neg\neg$- and $\neg\vee$-rules (together with regrouping and exchange, which we usually take for granted) suffice to simplify the Boolean part of a sentence. We can also assume that all quantifiers $\exists x$ have been rewritten in terms of $\forall x$, because

$$(\exists x)\mathcal{A}(x) \quad \text{is equivalent to} \quad \neg(\forall x)\neg\mathcal{A}(x).$$

It then remains to find falsification rules involving $\forall x$, and the two shown in Figure 4.4 suffice. As in Section 4.3, we associate falsification rules with downward directed edges of a tree. The aim is to falsify the upper sentence by falsifying the lower one.

It can be checked that an interpretation falsifies the upper sentence if and only if it falsifies the lower sentence. The reason for the restriction to

$(\forall x)\mathcal{A}(x) \vee \mathcal{B}$ $\neg(\forall x)\mathcal{A}(x) \vee \mathcal{B}$

$\quad\quad \Big| \forall$ $\quad\quad\quad \Big| \neg\forall$

$\mathcal{A}(a) \vee \mathcal{B}$ $\neg\mathcal{A}(a) \vee \neg(\forall x)\mathcal{A}(x) \vee \mathcal{B}$

for any constant a not free in $\mathcal{B}$ for any constant a

Figure 4.4. Quantifier falsification rules.

a not free in $\mathcal{B}$ in the rule on the left is to avoid getting the unfalsifiable sentence $\mathcal{A}(a) \vee \neg\mathcal{A}(a)$, which happens if $\mathcal{B} = \neg\mathcal{A}(a)$. The rule on the right does not really simplify the sentence, since the formula $\neg(\forall x)\mathcal{A}(x)$ persists, and hence some branches may not end.

However, if the construction of the tree is managed so that the constants $a_1, a_2, a_3, \ldots$ are introduced in order on each branch, and all applicable rules are applied, *any infinite branch will have the properties below*. To state these properties concisely we decompose each sentence in the tree as a disjunction, $\mathcal{A}_1 \vee \mathcal{A}_2 \vee \cdots \vee \mathcal{A}_k$, of the maximal number of sentences $\mathcal{A}_1, \mathcal{A}_2, \ldots, \mathcal{A}_k$ (called its *disjuncts*).

- If $\neg\neg\mathcal{A}$ occurs as a disjunct on the branch, so does $\mathcal{A}$.

- If $\neg(\mathcal{A}_1 \vee \mathcal{A}_2)$ occurs as a disjunct on the branch, so does $\neg\mathcal{A}_1$ or $\neg\mathcal{A}_2$.

- If $(\forall x)\mathcal{A}(x)$ occurs as a disjunct on the branch, so does $\mathcal{A}(a_i)$ for some constant a_i.

- If $\neg(\forall x)\mathcal{A}(x)$ occurs as a disjunct on the branch, so does $\neg\mathcal{A}(a_i)$ for *every* constant a_i.

It follows that, on an infinite branch, every sentence is ultimately broken down to *atoms*: terms of the form $R(a_i, a_j, \ldots)$ or $\neg R(a_i, a_j, \ldots)$ where R is a relation symbol.[2] Each atom occurs in only one form, either with the negation sign or without, because if both forms occur the disjunction containing them cannot be falsified and the branch will end. Thus we are free to assign each atom the value F, and *every sentence in the branch gets the value F as a result, including the sentence $\mathcal{S}$ at the root of the tree.*

Conversely, *if a sentence $\mathcal{S}$ of predicate logic is valid, every branch in the tree terminates, necessarily in a disjunction of terms that includes the same atom, both with and without a negation sign.*

Therefore, *by taking the sentences of the form $R(a_i, a_j, \ldots) \vee \neg R(a_i, a_j, \ldots)$ as axioms, and the reverses of the falsification rules as rules of inference, we get a complete formal system for predicate logic.*

The two new rules in predicate logic are the following.

$\forall$-**rule.** From $\mathcal{A}(x) \vee \mathcal{B}$ for a free variable x infer $(\forall x)\mathcal{A}(x) \vee \mathcal{B}$.

$\neg\forall$-**rule.** From $\quad \neg\mathcal{A}(a) \vee \mathcal{B} \quad$ infer $\quad \neg(\forall x)\mathcal{A}(x) \vee \mathcal{B}$.

This formal system does not contain a primitive rule that is the reverse of the second falsification rule, namely,

from $\neg\mathcal{A}(a) \vee \neg(\forall x)\mathcal{A}(x) \vee \mathcal{B}$ infer $\neg(\forall x)\mathcal{A}(x) \vee \mathcal{B}$.

---

[2]Admittedly, sentences of the form $\neg(\forall x)\mathcal{A}(x)$ are carried all the way down the branch, but they spawn disjuncts of the form $\neg\mathcal{A}(a_i)$, which can be broken down.

However, we can *derive* the latter rule as a combination of the $\neg\forall$ rule and consolidation,

from $\neg\mathcal{A}(a) \vee \neg(\forall x)\mathcal{A}(x) \vee \mathcal{B}$ infer $\neg(\forall x)\mathcal{A}(x) \vee \neg(\forall x)\mathcal{A}(x) \vee \mathcal{B}$
(by $\neg\forall$-rule),
from $\neg(\forall x)\mathcal{A}(x) \vee \neg(\forall x)\mathcal{A}(x) \vee \mathcal{B}$ infer $\neg(\forall x)\mathcal{A}(x) \vee \mathcal{B}$
(by consolidation).

It follows that if a sentence $\mathcal{S}$ cannot be falsified, then it has a proof in the system with the rules of propositional logic, plus the $\forall$- and $\neg\forall$-rules. This is (one version of) the *completeness theorem for predicate logic*.

Gödel's version used an axiom system for predicate logic essentially due to Frege in 1879, with the cut rule as one of the rules of inference. In 1934 Gentzen showed that cut could be eliminated, thus showing the completeness of cut-free systems. However, it was only in the 1950s that completeness proofs were given along the above lines. The idea of finding the necessary rules of inference by reversing a natural falsification procedure seems to have occurred to several logicians around 1955 (Beth, Hintikka, Lis, Smullyan, and van Heijenoort—see Annellis (1990)).

4.6 COMPLETENESS, CONSISTENCY, HAPPY ENDINGS

> If contradictory attributes be assigned to a concept, I say that mathematically the concept does not exist. ... But if it can be proved that the attributes assigned to the concept can never lead to a contradiction by the application of a finite number of logical inferences, I say that the mathematical existence of the concept ... is thereby proved.
>
> —David Hilbert
> Address on *Mathematical Problems*, 1900,
> in Ewald (1996), volume II, p. 1105.

The completeness proof for predicate logic is a remarkable vindication of Hilbert's concept of mathematical existence. Before Hilbert's time, mathematics had seen many examples of "imaginary" or "ideal" objects with no obvious reason to exist, other than their apparent consistency and usefulness.

A typical example is $\sqrt{-1}$, a mysteriously useful object that was used in calculations for several centuries before anyone understood what it meant. Perhaps the ultimate explanation, proposed by the French mathematician Augustin-Louis Cauchy in 1847, is that *the meaning of $\sqrt{-1}$ resides in its behavior as an algebraic symbol*. That is, we can consistently introduce a symbol i that obeys the rule $i^2 = -1$, and that is all we need to know. We are already used to the idea that it is consistent to calculate with a letter that obeys the same rules as numbers, and it can be proved that the same rules continue to hold when we impose the extra condition that $i^2 = -1$. (In modern algebraic language, "polynomials in i, mod $i^2 + 1$, form a field.")

The proof of the completeness theorem gives meaning to mathematical concepts via symbols in a much more general way. Briefly, the idea is this.

Suppose S is a sentence that does not lead to a contradiction. Then $\neg S$ can be falsified, because if $\neg S$ cannot be falsified we can *prove* $\neg S$, by the completeness theorem, and this certainly contradicts S. The falsification procedure gives a specific interpretation that falsifies $\neg S$, and hence makes S true. The interpretation in question assigns the values T or F to the relation symbols in S for certain constants. Thus the "concepts" implicit in S have meaning in a certain symbolic world, where the "individuals" $a, b, c, \ldots$ are constant symbols and relations $R(x, y)$ are determined by the truth values assigned to them when constants are substituted for their variables.

Hilbert's dream of existence for consistent concepts is therefore realized (at least for concepts expressible in the language of predicate logic) by the completeness theorem. We can say that Gödel made one of Hilbert's dreams come true, even though he crushed the other dream (of a finitary consistency proof for mathematics) with his incompleteness theorems for arithmetic.

Indeed, the incompleteness theorems add an interesting twist to Hilbert's concept of mathematical existence. Incompleteness of a particular axiom system $\mathcal{A}$ means that there is a sentence S such that *both* S and $\neg S$ are consistent with $\mathcal{A}$. The argument above then leads to *two* symbolic worlds: one in which $\mathcal{A}$ and S are true, and one in which $\mathcal{A}$ and $\neg S$ are true. Actually, this situation was already known long before Gödel, in the case of the *parallel axiom* in plane geometry. In the 19th century, geometers constructed various geometric worlds, some in which the parallel axiom is true, and some in which it is false.

PROOFS AND HAPPY ENDINGS

Obtaining proofs in predicate logic from failed falsification attempts has an element of uncertainty not present in propositional logic. Namely, the falsification procedure is potentially infinite, so we cannot in general tell how long it will take to prove a given sentence S, if S is in fact valid. This is unavoidable. If we always knew in advance how long to search for a proof, then we would have an algorithm for deciding validity, contrary to the unsolvability of the Entscheidungsproblem (Section 3.7). Thus we cannot always foresee how long falsification will take, or whether the process will be infinite.

On the other hand, since each valid sentence has a finite proof, it is possible to show that the formal system is sound by much the same evaluation argument that we used for propositional logic. Essentially one ignores quantifiers, evaluating all sentences T or F as in propositional logic, and showing that the value T propagates from axioms to theorems. Thus no false sentence is provable.

We can also show that each theorem of predicate logic has a "happy ending" proof in the following sense:

1. If the target theorem is of the form $\neg\neg\mathcal{A} \vee \mathcal{B}$, then the last step is

   from $\mathcal{A} \vee \mathcal{B}$ infer $\neg\neg\mathcal{A} \vee \mathcal{B}$.

2. If the target theorem is of the form $\neg(\mathcal{A}_1 \vee \mathcal{A}_2) \vee \mathcal{B}$, then the last step is

   from $\neg\mathcal{A}_1 \vee \mathcal{B}$ and $\neg\mathcal{A}_2 \vee \mathcal{B}$ infer $\neg(\mathcal{A}_1 \vee \mathcal{A}_2) \vee \mathcal{B}$.

3. If the target theorem is of the form $(\forall x)\mathcal{A}(x) \vee \mathcal{B}$, then the last step is

   from $\mathcal{A}(x) \vee \mathcal{B}$ infer $(\forall x)\mathcal{A}(x) \vee \mathcal{B}$.

4. If the target theorem is of the form $\neg(\forall x)\mathcal{A}(x) \vee \mathcal{B}$, then the last step is

   from $\neg\mathcal{A}(a) \vee \mathcal{B}$ infer $\neg(\forall x)\mathcal{A}(x) \vee \mathcal{B}$.

Cases 1 and 2 follow as they did for propositional logic in Section 4.3. Cases 3 and 4 are similar, using the falsification procedure for predicate logic.

For example, given a theorem of the form $\neg(\forall x)\mathcal{A}(x) \vee \mathcal{B}$, there must be some instance $\neg\mathcal{A}(a)$ for which $\neg\mathcal{A}(a) \vee \mathcal{B}$ is valid. Therefore, any attempt to falsify $\neg\mathcal{A}(a) \vee \mathcal{B}$ will fail, and running this attempt backwards yields a proof of $\neg\mathcal{A}(a) \vee \mathcal{B}$, from which $\neg(\forall x)\mathcal{A}(x) \vee \mathcal{B}$ follows immediately by the $\neg\forall$-rule.

4.7 Historical Background

Propositional Logic and Computation

...the mental work of a mathematician concerning Yes-or-No questions could be completely replaced by a machine.

—Kurt Gödel
Letter to John von Neumann, 20 March 1956,
in Gödel (2003), p. 375.

When Post published the truth table method in 1921, it appeared that all the basic questions about propositional logic had been settled. Truth tables decide the following.

- *Validity:* a formula is valid if and only if it has value "true" for all values of its variables (and the value of the formula is computable by truth tables).

- *Consistency:* the usual formal systems for propositional logic are consistent because their axioms have value "true," and the rules of inference preserve truth.

- *Satisfiability:* a formula is satisfiable if and only if it has value "true" for some assignment of truth values to its variables.

In particular, validity and satisfiability are testable in a finite time that can be bounded in advance. Since there are 2^n ways to assign truth values to n variables, the time required to test a formula with n variables is of the order of 2^n steps. In the early 20th century, this was as finite as anyone wanted. But in the 1950s, when the first computer programs for testing logical validity were developed, logicians began to realize that 2^n steps is a long time.

In the same period, Gödel, with typical prescience, wrote the following to John von Neumann: in the letter quoted above.

> I would like to allow myself to write you about a mathematical problem, of which your opinion would very much interest me: One can obviously easily construct a Turing machine, which for every formula F in first order predicate logic and every natural number n, allows one to decide if there is a proof of F of length n (length = number of symbols). Let $\psi(F, n)$ be the number of steps the machine requires for this and let $\varphi(n) = \max_F \psi(F, n)$. The question is how fast $\varphi(n)$ grows for an optimal machine. One can show that $\varphi(n) \geq k \cdot n$. If there really were a machine with $\varphi(n) \sim k \cdot n$ (or even $\sim k \cdot n^2$), this would have consequences of the greatest importance. Namely, it would obviously mean that in spite of the undecidability of the Entscheidungsproblem, the mental work of a mathematician concerning Yes-or-No questions could be completely replaced by a machine. After all, one would simply have to choose the natural number n so large that when the machine does not deliver a result, it makes no sense to think more about the problem.

Gödel is asking an apparently more general question—about deciding validity in predicate logic—but actually the issue is much the same as deciding satisfiability in propositional logic. In both cases one has to search through a space of size roughly 2^n candidates (sequences of n

predicate logic symbols, assignments of truth values to n propositional variables) for an object with an *easily recognizable* property:

- We can easily tell whether a given sequence of n symbols is a proof in predicate logic by breaking it into formulas and checking whether each follows from preceding formulas by one of the given rules of inference.

- We can easily tell whether a propositional formula in n variables is satisfied, for given values of the variables, by computing its value by truth tables.

These two tasks are "easy" in the sense that they can be done in around n^2 steps, in contrast to the approximately 2^n steps it takes to enumerate all objects in the space of candidates. Thus the "hard" part—the part for which we seem to need mathematicians—is finding a validating object in an exponentially large space.

When the two problems are compared in this way, it looks as though *the satisfiability problem of propositional logic is all mathematics, in microcosm.* In a sense, this is correct. Gödel's problem of deciding, for given n, whether a given sentence has a proof of length at most n reduces to deciding satisfiability of a propositional formula with roughly n variables. "Roughly" means "bounded by some polynomial function of n." This follows from a more general result of Stephen Cook (1971) about *nondeterministic Turing machines.*

A nondeterministic Turing machine is the same as an ordinary Turing machine (described in Section 3.5), except that a given (state, symbol) pair (q_i, S_j) may occur in *more* than one quintuple. Thus, more than one computation is possible on a given input tape. This of course is not how we want real machines to behave, but it is great for a hypothetical machine, because it allows the machine to *make a guess and then test it.* For example, suppose we include the quintuples

$$q_1 B0Rq_1, \quad q_1 B1Rq_1, \quad q_1 BBRq_2$$

in the machine, where B denotes the blank square of tape. Then, when started in state q_1 on a blank tape, the machine can write a random sequence of 0s and 1s and jump into state q_2. Once in state q_2, the machine can treat the sequence of 0s and 1s as an assignment of truth values to a propositional formula $\mathcal{F}$ given as input (say, to the left of the starting square). By implementing the truth table method, the machine can compute the truth value of $\mathcal{F}$ for the sequence of values of the variables that it has randomly generated.

Clearly, if the propositional formula $\mathcal{F}$ of length n is satisfiable, then this fact is shown by a nondeterministic Turing machine computation of roughly n steps. By a similar argument, if a sentence of predicate logic has a proof of length at most n, this fact is also shown by a nondeterministic Turing machine computation of roughly n steps. Cook proved the following (it was also discovered independently by Leonid Levin) by describing the actions of a Turing machine in terms of "and," "or," and "not":

COOK'S THEOREM. *For each nondeterministic Turing machine M, input I, and time T, there is a propositional formula $\mathcal{F}_{M,I,T}$, of length roughly the size of M, I, T, such that:*

M *halts, for input I, in at most T steps if and only if $\mathcal{F}_{M,I,T}$ is satisfiable.*

As I said, "roughly the size of" means "bounded by a polynomial function of." So the theorem is about the so-called class NP of problems "solvable nondeterministically in polynomial time." A problem $\mathcal{P}$ is in NP if there is a nondeterministic Turing machine M and a polynomial p such that M answers each question in $\mathcal{P}$ of length n by *some* computation of at most $p(n)$ steps. Cook's theorem says that any problem in NP is "easy," relative to the satisfiability problem. If (by some miracle) there is a *deterministic* polynomial-time solution of the satisfiability problem, then there is a deterministic polynomial-time solution of every problem in NP.

The problems solvable deterministically in polynomial time make up a class called P. In his letter to von Neumann, Gödel asked about deterministic solution in kn or kn^2 steps. But if we relax his step bound to an arbitrary polynomial $p(n)$, then he is asking whether class P includes the problem of deciding the existence of proofs of length n for formulas of predicate logic. The latter problem includes the satisfiability problem, so Gödel is implicitly asking whether NP equals P. This has become the most important open question in logic (and computer science) today. As he says, it is the question of whether mathematicians can be replaced by machines.

BETWEEN PROPOSITIONAL LOGIC AND PREDICATE LOGIC

I am still *perfectly convinced* that reluctance to use non-finitary concepts ... was the primary reason why the completeness theorem was not given by ... anyone else before my work

—Kurt Gödel
Letter to Hao Wang, 7 March 1968,
in Gödel (2003), p. 403.

In the completeness proof for predicate logic in Section 4.5 there is a mild assumption about infinity. We assumed that the proof tree, if it is not finite, has an infinite branch. This assumption is a special case of the following result, which was published by the Hungarian mathematician Dénes König in 1926 and made more widely known by his book König (1936/1990):

THE KÖNIG INFINITY LEMMA

If T is an infinite tree in which each vertex meets only finitely many edges, then T contains an infinite branch.

Start at any vertex in T, and consider its finitely many outgoing edges. At least one of these edges, e, leads into a subtree T' which is also infinite. So, move to the vertex at the other end of e and repeat the argument. Continuing the process indefinitely, we obtain an infinite branch.

The König infinity lemma has some infinite content which disqualifies it from the Hilbert program of finitary proofs. For example, it is possible to define an infinite computable tree with no computable infinite branch. In fact, Gödel believed that he succeeded in proving the completeness of predicate logic, where others had failed, because of his willingness to step outside the Hilbert program. (Bear in mind that Gödel's completeness proof for predicate logic came a year before his incompleteness theorems effectively ended the Hilbert program.)

We now know that the König infinity lemma is needed for any proof that predicate logic is complete. A whole subject, called *reverse mathematics*, has sprung up to study the assumptions on which key theorems depend. The König infinity lemma also underlies some classical theorems of analysis, such as the Heine-Borel theorem stating that any covering of $[0, 1]$ by open intervals includes a finite subcover. For more on this, see Simpson (1999).

König (1936/1990) said that his lemma "furnishes a useful method of carrying over certain results from the finite to the infinite," and illustrated the method by the Heine-Borel theorem and some theorems of combinatorics. Today, the property of $[0, 1]$ expressed by the Heine-Borel theorem is called *compactness*, and other properties derived from the König infinity lemma are also called "compactness." An important example is:

COMPACTNESS THEOREM FOR PROPOSITIONAL LOGIC. *If S is an infinite set of formulas of propositional logic, any finite subset of which is satisfiable, then the whole set S is satisfiable.*

Proof: To launch the proof of this theorem, we let $P_1, P_2, P_3, \ldots$ be all the propositional variables occurring in formulas in S, and consider the *tree of all assignments of truth values to $P_1, P_2, P_3, \ldots$*. By renaming variables, if

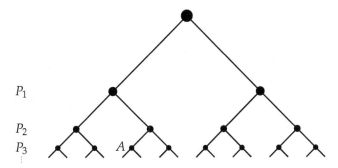

Figure 4.5. The tree of truth-value assignments.

necessary, we can ensure that there are infinitely many of them, so this is an infinite tree. The first few levels are shown in Figure 4.5. Each of the 2^n vertices at level n represents an assignment of values to $P_1, P_2, \ldots, P_n$. For example, the vertex A at the end of the branch that goes left, right, left represents the assignment

$$P_1 = \text{false}, \quad P_2 = \text{true}, \quad P_3 = \text{false}.$$

Now we *prune* this tree by deleting each vertex for which the corresponding assignment falsifies some formula in S. What remains is still an infinite tree, T. If not, then all branches terminate by some level n. In that case, each assignment A of truth values to $P_1, P_2, \ldots, P_n$ falsifies some formula F_A in S. But then the finite set of formulas F_A is unsatisfiable, contrary to assumption.

Thus, by the König infinity lemma, T has an infinite branch. The values assigned to $P_1, P_2, P_3, \ldots$ along this branch necessarily satisfy all formulas in S, otherwise the branch would not have survived the pruning process. $\square$

The compactness theorem for propositional logic has a certain smell of predicate logic about it, and not only in the style of proof. It really *implies* the corresponding compactness theorem for predicate logic, thanks to a trick that replaces each formula of predicate logic by a (usually infinite) set of formulas of propositional logic.

The compactness theorem for predicate logic was actually discovered first. It was one of the steps in Gödel's proof of the completeness theorem in 1929. As is often the case, its essence—in propositional logic and ultimately in the König infinity lemma—was distilled out only later.

- CHAPTER 5 -

ARITHMETIC

PREVIEW

In the previous chapter we showed that logic is "simple" in the sense that its valid sentences can be proved in a straightforward way. Proofs proceed by natural rules that build up the target sentence from simpler sentences, essentially by reversing an exhaustive attempt to falsify the target sentence by breaking it down to simpler parts.

The simplicity of logic without cuts allows its consistency to be proved by elementary means, available in formal arithmetic. The consistency of arithmetic itself *cannot* be proved by these means—by Gödel's second theorem—but a natural extension of them suffices.

The essential ingredient of formal arithmetic is induction over the positive integers, that is, ordinals less than ω. Induction over finite ordinals suffices to prove the consistency of logic, basically because logically valid sentences can be recognized by finite processes. Truth in arithmetic involves consideration of infinite processes, since arithmetic contains induction over finite ordinals. Induction also forces us to use the cut rule. As a result, the logic of arithmetic is quite complex, and its consistency can only be proved only by *transfinite* induction. The appropriate induction goes up to the ordinal ε_0.

We outline a consistency proof for arithmetic by describing proofs in a standard axiom system PA ("Peano arithmetic") and their translations into a system PA_ω, which is in some ways simpler. The catch is that proofs in PA_ω may be infinite, and it is necessary to eliminate cuts from them before they become simple enough to ensure the consistency of PA. This involves assigning ordinals $< \varepsilon_0$ to proofs in PA_ω, and using ε_0-induction to eliminate cuts.

5.1   HOW MIGHT WE PROVE CONSISTENCY?

In the sections below we will give a formal system for "arithmetic" or "elementary number theory," and outline a proof that it is consistent. As we know from Gödel's second incompleteness theorem, such a proof must involve an axiom that goes beyond arithmetic itself. And as we have intimated in Section 3.4, a natural axiom to choose is ε_0-induction, where ε_0 is the countable ordinal introduced in Section 2.1. But before we explain why ε_0 is the right ordinal to describe the complexity of arithmetic, let us hear from Gerhard Gentzen, who discovered the connection between ordinals and consistency. The following passage is from a piece he wrote for a general audience in 1936, *Der Unendlichkeitsbegriff in der Mathematik* (The concept of infinity in mathematics), shortly after the publication of his consistency proof. (See Gentzen (1969), pp. 231–232.)

> I shall explain now how the concepts of the transfinite ordinal numbers and the rule of transfinite induction enter into the consistency proof. The connection is quite natural and simple. In carrying out a consistency proof for elementary number theory we must consider all conceivable number-theoretical proofs and must show that in a certain sense, to be formally defined, each individual proof yields a 'correct' result, in particular, no contradiction. The 'correctness' of the proof depends on the correctness of certain other simpler proofs contained in it as special cases or constituent parts. This fact has motivated the arrangement of proofs in linear *order* in such a way that those proofs on whose correctness the correctness of another depends precede the latter proof in the sequence. This arrangement of proofs is brought about by correlating with each proof a certain transfinite ordinal number; the proofs preceding a given proof are precisely those proofs whose ordinal numbers precede the ordinal number of the given proof ... we need the transfinite ordinal numbers for the following reason: It may happen that the correctness of a proof depends on the correctness of infinitely many simpler proofs. An example: Suppose that in the proof a proposition is proved for *all* natural numbers by complete induction. In that case the correctness of the proof obviously depends on the correctness of every single one of the infinitely many individual proofs obtained by specializing to a particular natural number. Here a natural number is insufficient as an ordinal number for the proof, since each natural number is preceded by only finitely many other numbers in the natural ordering. We therefore need the transfinite ordinal numbers in order to represent the natural ordering of the proofs according to their complexity.

Gentzen faithfully describes the main ideas of his proof, but there are two claims that may seem puzzling to anyone with some experience of proofs in number theory.

1. The claim that any proof may be based on "simpler proofs contained in it." It often happens in mathematics that a simple conclusion is drawn from complicated premises. This is typically due to the modus ponens, or cut rule (as Gentzen called it): from $P \Rightarrow Q$ and P, infer Q. The conclusion Q contains no trace of the (possibly more complicated) proposition P, yet it seems to depend on it. Gentzen got around this problem in an earlier work (and we did in Chapter 4) by showing that any proof in standard logic may be reorganized to *eliminate all cuts*.

2. The claim that a proof by induction depends on the "infinitely many proofs obtained by specializing to particular natural numbers." In proving a sentence $\mathcal{S}(n)$ about a positive integer n one does not literally prove the infinitely many propositions $\mathcal{S}(1), \mathcal{S}(2), \mathcal{S}(3), \ldots$. Instead, one proves the special proposition $\mathcal{S}(1)$ and then the *general* proposition "$\mathcal{S}(k) \Rightarrow \mathcal{S}(k+1)$ for all k." Nevertheless, the ordinal level of the general proposition "$\mathcal{S}(n)$ holds for all n" is naturally placed above the levels of all the special propositions $\mathcal{S}(1), \mathcal{S}(2), \mathcal{S}(3), \ldots$.

These two claims point to two separate issues in the consistency proof for elementary number theory: the issue of cut elimination, which already arises in pure logic, and the issue of induction, which is specific to number theory. Since we have given a cut-free approach to logic in Sections 4.3 and 4.5, it remains to discuss induction.

5.2 FORMAL ARITHMETIC

We take the language of arithmetic to be that in Section 4.4, except that we drop the relation $m < n$, since it can be defined by

$$(\exists l)(l + m = n).$$

(This assumes that our variables range over *positive* integers, so the definitions of plus and times from Section 2.5 will be adjusted accordingly.) The appropriate logic for this language is predicate logic. It would be nice if we could use the rules of inference for predicate logic, and take the axioms of arithmetic to be obvious facts about numbers, such as 1+1=2. This idea does not quite work, however, because induction is crucial. To prove a sentence $(\forall n)\mathcal{P}(n)$, where $\mathcal{P}$ denotes some property written in our formal language, we generally have to prove the sentences $\mathcal{P}(1)$ and $(\forall n)(\mathcal{P}(n) \Rightarrow \mathcal{P}(n+1))$.

This amounts to using infinitely many induction axioms of the form

$$[\mathcal{P}(1) \wedge (\forall n)(\mathcal{P}(n) \Rightarrow \mathcal{P}(n+1))] \Rightarrow (\forall n)\mathcal{P}(n), \qquad (n \to n+1)$$

one for each property $\mathcal{P}$ expressible in the language of arithmetic. Almost any general sentence about positive integers depends on induction, so one may need a complicated axiom to prove a simple sentence like $(\forall n)(1 + n = n + 1)$. Typically, a complicated premise A yields a simple conclusion by the cut rule,

$$\text{from} \quad A \Rightarrow B \quad \text{and} \quad A \quad \text{infer} \quad B.$$

Indeed Gentzen's consistency proof for arithmetic is based on *eliminating cuts* from a formal system with the cut rule, an idea he had introduced two years earlier for predicate logic. However, we now know that all valid sentences of predicate logic can be proved without even thinking about the cut rule—as in Section 4.4—so it is tempting to try to do the same for arithmetic.

One way to avoid cuts and the complicated induction axioms is to use an idea introduced by the German mathematician Kurt Schütte in 1951. We "embed" ordinary arithmetic in a system with the so-called ω-*rule*,

$$\text{from} \quad \mathcal{P}(1), \mathcal{P}(2), \mathcal{P}(3), \ldots \quad \text{infer} \quad (\forall n)\mathcal{P}(n). \qquad (\omega)$$

This rule, with its infinitely many premises, may seem like cheating—because a human being (or a computer) cannot actually execute it. However, the logical consequences of the ω-rule are a mathematically well-defined set, which we can prove *contains only true sentences of arithmetic*. Therefore, if we prove that all consequences of the induction rules ($n \to n + 1$) can be obtained by the ω-rule, it will follow that formal arithmetic is also consistent.

The catch is that, at first sight, the only way to carry over induction proofs is to include the cut rule along with the ω-rule. We then have to show that all cuts can be eliminated, thus getting back to the consistent cut-free system.

In Section 5.4 we sketch how any consequence of the ordinary induction axioms ($n \to n + 1$) can be derived using the ω-rule. Finally, in Section 5.5 we eliminate cuts from the ω-rule system by transfinite induction. It turns out to be necessary and sufficient to use induction up to the ordinal ε_0.

5.3   The Systems PA and PA$_\omega$

To explain how the formal system of ordinary arithmetic can be embedded in a system with the ω-rule, we need to describe both systems more precisely. Ordinary arithmetic is a system commonly known as *Peano arithmetic*, and denoted by PA.

Its axioms get their name from the Italian mathematician Giuseppe Peano, who stated them in 1889. Essentially the same axioms had already been studied by Dedekind, and Peano acknowledged the influence of both Dedekind and Grassmann on his work. All three mathematicians recognized the crucial role of induction, which is usually expressed today in the form $(n \to n + 1)$ given in the previous section.[1] Thus, the following is what we usually mean by "Peano Arithmetic" today.

AXIOMS OF PA

The axioms of PA state some obvious properties of the positive integers. As in ordinary mathematics, we abbreviate the inequation $\neg(m = n)$ by $m \neq n$, and interpret statements with free variables $m, n, \ldots$ as true for all m, all n, ...

1. The number 1 is not a successor: $1 \neq S(n)$.

2. Numbers with the same successor are equal: $S(m) = S(n) \Rightarrow m = n$

3. Inductive definition of sum:

$$m + 1 = S(m), \quad m + S(n) = S(m + n).$$

4. Inductive definition of product:

$$m \cdot 1 = m, \quad m \cdot S(n) = m \cdot n + m.$$

5. Induction: for all properties $\mathcal{P}$ expressible in the language,

$$[\mathcal{P}(1) \wedge (\forall n)(\mathcal{P}(n) \Rightarrow \mathcal{P}(n + 1))] \Rightarrow (\forall n)\mathcal{P}(n), \qquad (n \to n + 1)$$

(Since there are infinitely many instances of $\mathcal{P}$, this is more properly called an *axiom schema*.)

---

[1]Peano and Dedekind stated induction $(n \to n + 1)$ for *arbitrary* properties $\mathcal{P}$ of the positive integers. As we know from Chapter 1, there are uncountably many such properties. This is more than can be described in our language for arithmetic, because our language (and indeed, any language readable by humans) contains only countably many strings of symbols. This makes a completely general statement of induction problematic, and it is usual to require induction only for properties that can be defined in a reasonably expressive language. After all, we wish to prove only sentences expressible in PA, so we should need only instances of induction expressible in PA.

Rules of inference of PA

As rules of inference we take the rules for predicate logic given in Sections 4.3 and 4.4, together with the *cut rule*:

Cut. From $\mathcal{B} \vee \mathcal{A}$ and $\neg \mathcal{A} \vee \mathcal{C}$ infer $\mathcal{B} \vee \mathcal{C}$.

The cut rule is essentially a symmetric form of modus ponens, as one sees by translating $\neg \mathcal{A} \vee \mathcal{C}$ into $\mathcal{A} \Rightarrow \mathcal{C}$. Modus ponens says that from $\mathcal{A}$ and $\mathcal{A} \Rightarrow \mathcal{C}$ we can infer $\mathcal{C}$. Therefore, from $\mathcal{B} \vee \mathcal{A}$ and $\mathcal{A} \Rightarrow \mathcal{C}$ we can infer $\mathcal{B} \vee \mathcal{C}$.

Arithmetic with the ω-rule is a system with very simple axioms, and rules of inference like those of predicate logic, except that the ω-rule replaces the usual $\forall$-rule. We call the system PA_ω because it turns out to prove all theorems of PA. As mentioned in the previous section, the cut rule is in PA_ω only to facilitate the transfer of proofs from PA. PA_ω *minus* the cut rule is in fact a complete system; its theorems are precisely the true sentences of PA.

Axioms of PA_ω

The axioms of PA_ω are basically what we called "arithmetic" in elementary school, namely, all true facts about sums and products of numbers. To be precise, we take as axioms all true equations involving only constants (such as $1 + 1 = 2$), and the negations of all false equations involving only constants (such as $2 \cdot 2 \neq 3$).

Rules of inference of PA_ω

As rules of inference for PA_ω we take the cut rule, the propositional rules of Section 4.3, the $\neg\forall$-rule for predicate logic of Section 4.5, and (in place of the $\forall$-rule) the ω-rule of Section 5.2.

5.4   Embedding PA in PA_ω

PA_ω includes all the rules of inference of propositional logic, but its axioms do not include the logic axioms $\mathcal{A} \vee \neg \mathcal{A}$. So it is not immediately clear that PA_ω proves all sentences of PA valid in propositional logic.

To show that any sentence of the form $\mathcal{A} \vee \neg \mathcal{A}$ is a theorem of PA_ω we use induction on the number of logical symbols in $\mathcal{A}$ (called the *degree* of $\mathcal{A}$).

Base step. If $\mathcal{A}$ has no logic symbols then either $\mathcal{A}$ or $\neg \mathcal{A}$ is an axiom of PA_ω, and hence $\mathcal{A} \vee \neg \mathcal{A}$ follows by the dilution rule.

INDUCTION STEP. If $\mathcal{A}$ has logic symbols then it has the form $\neg\mathcal{A}_1$, $\mathcal{A}_1 \vee \mathcal{A}_2$, or $(\forall x)\mathcal{A}(x)$, where $\mathcal{A}_1$, $\mathcal{A}_2$, and $\mathcal{A}(x)$ have lower degree. We can therefore assume, by induction, that $\mathcal{A}_1 \vee \neg\mathcal{A}_1$ and the like are theorems of PA$_\omega$. We can then use rules of PA$_\omega$ to prove $\mathcal{A} \vee \neg\mathcal{A}$.

For example, if we assume that $\mathcal{A}(x) \vee \neg\mathcal{A}(x)$ is a theorem for any free variable x, then we can prove

$$\mathcal{A}(1) \vee \neg\mathcal{A}(1), \quad \mathcal{A}(2) \vee \neg\mathcal{A}(2), \quad \ldots$$

by replacing x in the proof by $1, 2, \ldots$. Next we can prove

$$\mathcal{A}(1) \vee \neg(\forall x)\mathcal{A}(x), \quad \mathcal{A}(2) \vee \neg(\forall x)\mathcal{A}(x), \quad \ldots$$

by the $\neg\forall$-rule, and finally $(\forall x)\mathcal{A}(x) \vee \neg(\forall x)\mathcal{A}(x)$ follows by the ω-rule.

Thus PA$_\omega$ proves all sentences of PA valid in propositional logic.

It follows that PA$_\omega$ also proves all sentences of PA valid in predicate logic. PA$_\omega$ includes the $\neg\forall$ rule of predicate logic, and we can effect applications of the $\forall$ rule by applications of its replacement in PA$_\omega$, the ω-rule. Namely, if we can prove $\mathcal{A}(x)$ for a free variable x, then we can also prove $\mathcal{A}(1), \mathcal{A}(2), \ldots$, simply by replacing the letter x in the proof by $1, 2, \ldots$. And then $(\forall x)\mathcal{A}(x)$, the consequence of $\mathcal{A}(x)$ by the $\forall$ rule, follows by the ω-rule.

How do we prove the axioms of PA in PA$_\omega$?

The first four axioms of PA are easily proved in PA$_\omega$, thanks to the ω-rule. For example, to prove the first axiom, $(\forall n)[1 \neq S(n)]$, it suffices to apply the ω-rule to the infinitely many sentences

$$1 \neq S(1), \quad 1 \neq S(2), \quad 1 \neq S(3), \quad \ldots$$

all of which are axioms of PA$_\omega$.

The most difficult case is the induction axiom

$$[\mathcal{P}(1) \wedge (\forall n)(\mathcal{P}(n) \Rightarrow \mathcal{P}(n+1))] \Rightarrow (\forall n)\mathcal{P}(n), \qquad (n \rightarrow n+1)$$

which is logically equivalent to

$$\mathcal{P}(1) \Rightarrow [(\forall n)(\mathcal{P}(n) \Rightarrow \mathcal{P}(n+1)) \Rightarrow (\forall n)\mathcal{P}(n)],$$

and hence can be written in terms of $\neg, \vee, \forall$ as

$$\neg\mathcal{P}(1) \vee \neg(\forall n)(\neg\mathcal{P}(n) \vee \mathcal{P}(n+1)) \vee (\forall n)\mathcal{P}(n)$$

To prove this statement of induction in PA_ω, first notice that the following sentences are valid in propositional logic (and hence provable in PA_ω).

$\neg P(1) \vee P(1),$

$\neg P(1) \vee \neg[\neg P(1) \vee P(2)] \vee P(2),$

$\vdots$

$\neg P(1) \vee \neg[\neg P(1) \vee P(2)] \vee \cdots \vee \neg[\neg P(k) \vee P(k+1)] \vee P(k+1).$

(The first is obvious, and each of the rest follows easily from its predecessor, particularly if one translates $\neg[\neg P(i) \vee P(i+1)]$ into $P(i) \wedge \neg P(i+1)$.)

Each bracketed term has the form

$$\neg[\neg P(i) \vee P(i+1)],$$

from which we get the same consequence

$$\neg(\forall n)[\neg P(n) \vee P(n+1)]$$

by the $\neg\forall$-rule. Consolidating the latter identical terms, we get the theorem

$$\neg P(1) \vee \neg(\forall n)[\neg P(n) \vee P(n+1)] \vee P(k+1).$$

We can prove this theorem for $k+1 = 2, 3, \ldots$, and in fact also for $k+1 = 1$ (using the theorem $\neg P(1) \vee P(1)$ and dilution), so we can apply the ω-rule to $P(k+1) = P(1), P(2), \ldots$ and get

$$\neg P(1) \vee \neg(\forall n)[\neg P(n) \vee P(n+1)] \vee (\forall n)P(n)$$

which is the induction axiom.

Thus *all the axioms of PA are theorems of PA_ω*.

Any *theorem* S of PA is a logical consequence of finitely many axioms, say $A_1, A_2, \ldots, A_k$, in which case

$$A_1 \wedge A_2 \wedge \cdots \wedge A_k \Rightarrow S, \quad \text{that is,} \quad (\neg A_1) \vee (\neg A_2) \vee \cdots \vee (\neg A_k) \vee S$$

is logically valid. It follows, by the completeness of predicate logic, that the sentence $A_1 \wedge A_2 \wedge \cdots \wedge A_k \Rightarrow S$ can be proved by the rules for predicate logic, and hence it can be proved in PA_ω, by the remarks above.

Finally, given that $A_1, A_2, \ldots, A_k$ are axioms of PA, hence theorems of PA_ω, and $A_1 \wedge A_2 \wedge \cdots \wedge A_k \Rightarrow S$ is also a theorem, we get S as a theorem of PA_ω by the cut rule. Thus *every theorem of PA is a theorem of PA_ω*.

In PA_ω it is in fact possible to prove all theorems *without* using the cut rule, which gives an easy proof that PA_ω is consistent. However, as one can guess from the embedding of PA in PA$_\omega$ carried out above, it is hard to see why the theorems of PA are theorems of PA$_\omega$ without using cut. This is why we include cut among the rules of PA$_\omega$ in the beginning, and only later show that all theorems of PA$_\omega$ can be proved without it. This process—*cut elimination* for PA$_\omega$—will be sketched in the next section. Cut elimination is where transfinite induction and the ordinal ε_0 come into the story.

5.5 CUT ELIMINATION IN PA$_\omega$

Proofs in PA, like the formal proofs described in Sections 4.3 and 4.5, can be viewed as finite trees. The branches begin with axioms and lead towards the theorem at the root. In showing that theorems of PA are theorems of PA$_\omega$ in Section 5.4 we essentially performed "tree surgery," converting proof-trees in PA to proof-trees in PA$_\omega$. In the surgery, the tree may become infinite, due to use of the ω-rule in building instances of the induction axioms. However, the proof-tree in PA$_\omega$ has only *finite branches*,[2] and *this allows us to assign a countable ordinal to each vertex in the tree as follows.*

Imagine the tree positioned so that the ends of branches lie at the top and the root lies at the bottom. Thus inference proceeds "downwards," with the conclusion of a rule of inference at the vertex "below" the vertices of the premises. We assign the ordinal 1 to each axiom at the end of a branch, and to each conclusion we assign the least ordinal greater than the ordinals of the premises. This procedure in fact assigns countable ordinals uniquely to the vertices of *any* countable tree with finite branches. Figure 5.1 shows an example. All the branches are finite but, because they are unbounded in length, their common vertex gets the ordinal ω.

Arbitrary countable trees with only finite branches may contain arbitrary countable ordinals as labels. Proof-trees in PA$_\omega$ are subject to certain constraints that keep their ordinals below ε_0, but we will not try to bound their ordinals yet. Our first goal is to show that every theorem of PA$_\omega$ has a cut-free proof, which we do as follows.

For each theorem $\mathcal{S}$ of PA$_\omega$, let the *ordinal of $\mathcal{S}$* be the least ordinal labeling $\mathcal{S}$ in all proof-trees of PA$_\omega$. If there are any theorems of PA$_\omega$ with no cut-free proof, let $\mathcal{T}$ be a theorem among them with least ordinal. Suppose that, in some proof of $\mathcal{T}$ with minimal ordinal, $\mathcal{T}$ is inferred from the sentences $\mathcal{U}_1, \mathcal{U}_2, \ldots$. Then $\mathcal{U}_1, \mathcal{U}_2, \ldots$ have ordinals smaller than the ordinal of $\mathcal{T}$, and hence they have cut-free proofs.

---

[2]This does not contradict the König infinity lemma, mentioned in Section 4.7, because in a proof tree there may be vertices that meet an infinite number of edges.

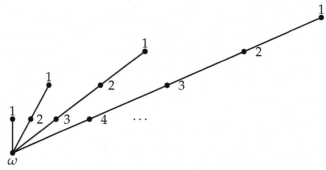

Figure 5.1. Ordinals labelling a tree with only finite branches.

By putting cut-free proofs of $\mathcal{U}_1, \mathcal{U}_2, \ldots$ together, we obtain a proof of $\mathcal{T}$ in which the inference of $\mathcal{T}$ from $\mathcal{U}_1, \mathcal{U}_2, \ldots$ is necessarily the *only* cut. Thus in fact $\mathcal{T}$ is inferred from $\mathcal{U}_1$ and $\mathcal{U}_2$ by the cut rule, following cut-free proofs of $\mathcal{U}_1$ and $\mathcal{U}_2$. Therefore, if this last cut can be eliminated we have a cut-free proof of $\mathcal{T}$—a contradiction—so we can conclude that all theorems of PA_ω have cut-free proofs.

How to Eliminate a Cut

This involves another induction, this time on the so-called *degree* of the cut,

from $\mathcal{B} \vee \mathcal{A}$ and $\neg \mathcal{A} \vee \mathcal{C}$ infer $\mathcal{B} \vee \mathcal{C}$,

which is defined to be the degree of the sentence $\neg \mathcal{A}$, that is, the number of logic symbols in $\neg \mathcal{A}$.

Suppose that we have a proof of a theorem $\mathcal{T}$ in which a cut occurs only at the very last step. We now show how to eliminate this cut altogether by replacing it by cuts of lower degree (which can then be replaced by cuts of still lower degree, and so on, until all cuts vanish). There are three cases, according as $\mathcal{A}$ has the form $\neg \mathcal{A}_1$, $\mathcal{A}_1 \vee \mathcal{A}_2$, or $(\forall x)\mathcal{A}(x)$.

1. When $\mathcal{A} = \neg \mathcal{A}_1$ the cut is

from $\mathcal{B} \vee \neg \mathcal{A}_1$ and $\neg\neg \mathcal{A}_1 \vee \mathcal{C}$ infer $\mathcal{B} \vee \mathcal{C}$,

and this is the last step in a proof that contains cut-free subproofs of $\mathcal{B} \vee \neg \mathcal{A}_1$ and $\neg\neg \mathcal{A}_1 \vee \mathcal{C}$. We know from the "happy ending" remarks of Section 4.3 that the cut-free proof of $\neg\neg \mathcal{A}_1 \vee \mathcal{C}$ can be arranged so that *its* last step is

from $\mathcal{A}_1 \vee \mathcal{C}$ infer $\neg\neg \mathcal{A}_1 \vee \mathcal{C}$.

But then we can infer our theorem $\mathcal{T} = \mathcal{B} \vee \mathcal{C}$ by the lower-degree cut

from $\mathcal{B} \vee \neg \mathcal{A}_1$ and $\mathcal{A}_1 \vee \mathcal{C}$ infer $\mathcal{B} \vee \mathcal{C}$,

where $\mathcal{A}_1 \vee \mathcal{C}$ is the penultimate sentence in the proof of $\neg \neg \mathcal{A}_1 \vee \mathcal{C}$.

2. When $\mathcal{A} = \mathcal{A}_1 \vee \mathcal{A}_2$ the cut is

from $\mathcal{B} \vee (\mathcal{A}_1 \vee \mathcal{A}_2)$ and $\neg (\mathcal{A}_1 \vee \mathcal{A}_2) \vee \mathcal{C}$ infer $\mathcal{B} \vee \mathcal{C}$,

and this is the last step in a proof that contains cut-free subproofs of $\mathcal{B} \vee (\mathcal{A}_1 \vee \mathcal{A}_2)$ and $\neg (\mathcal{A}_1 \vee \mathcal{A}_2) \vee \mathcal{C}$. Again we know, from the "happy ending" remarks of Section 4.3, that the cut-free proof of $\neg (\mathcal{A}_1 \vee \mathcal{A}_2) \vee \mathcal{C}$ can be arranged so that *its* last step is

from $\neg \mathcal{A}_1 \vee \mathcal{C}$ and $\neg \mathcal{A}_2 \vee \mathcal{C}$ infer $\neg (\mathcal{A}_1 \vee \mathcal{A}_2) \vee \mathcal{C}$.

But then we can infer our theorem $\mathcal{T} = \mathcal{B} \vee \mathcal{C}$ from two lower-degree cuts:

from $\mathcal{B} \vee \mathcal{A}_1 \vee \mathcal{A}_2$ and $\neg \mathcal{A}_1 \vee \mathcal{C}$ infer $\mathcal{B} \vee \mathcal{A}_2 \vee \mathcal{C}$,

from $\mathcal{B} \vee \mathcal{A}_2 \vee \mathcal{C}$ and $\neg \mathcal{A}_2 \vee \mathcal{C}$ infer $\mathcal{B} \vee \mathcal{C}$.

3. When $\mathcal{A} = (\forall x)\mathcal{A}(x)$ the cut is

from $\mathcal{B} \vee (\forall x)\mathcal{A}(x)$ and $\neg (\forall x)\mathcal{A}(x) \vee \mathcal{C}$ infer $\mathcal{B} \vee \mathcal{C}$,

and this is the last step in a proof that contains cut-free subproofs of $\mathcal{B} \vee (\forall x)\mathcal{A}(x)$ and $\neg (\forall x)\mathcal{A}(x) \vee \mathcal{C}$. Yet again we know, this time from the "happy ending" remarks of Section 4.6, that the cut-free proof of $\neg (\forall x)\mathcal{A}(x) \vee \mathcal{C}$ can be arranged so that its last step is

from $\neg \mathcal{A}(a) \vee \mathcal{C}$ infer $\neg (\forall x)\mathcal{A}(x) \vee \mathcal{C}$,

for some positive integer a. But we also have, by a similar "happy ending" argument, a cut-free proof of $\mathcal{B} \vee (\forall x)\mathcal{A}(x)$ in which the last step derives it from all the sentences in which $(\forall x)\mathcal{A}(x)$ is replaced by $\mathcal{A}(1), \mathcal{A}(2), \mathcal{A}(3), \ldots$. These include $\mathcal{B} \vee \mathcal{A}(a)$, so we can obtain our theorem $\mathcal{T} = \mathcal{B} \vee \mathcal{C}$ from the lower-degree cut

from $\mathcal{B} \vee \mathcal{A}(a)$ and $\neg \mathcal{A}(a) \vee \mathcal{C}$ infer $\mathcal{B} \vee \mathcal{C}$.

So in all cases we can replace the single cut in the proof of $\mathcal{T}$ by one or two cuts of lower degree. The reader may feel momentary anxiety that the number of cuts may grow beyond all bounds as the degree is reduced, but this is not the case if one always replaces the first cut in the tree. Thus $\mathcal{T}$ has a cut-free proof, and hence so do all theorems of PA$_\omega$.

CONSISTENCY OF PA$_\omega$ AND PA

One of the basic principles of logic, provable in propositional logic, is that a contradiction implies any proposition. Thus, if PA$_\omega$ proves any contradiction, it also proves the false proposition $1 = 2$. But if $1 = 2$ has a proof in PA$_\omega$ it has a cut-free proof, as we have just shown. And it is clear from the non-cut rules of PA$_\omega$ that $1 = 2$ can appear in the conclusion of a cut-free proof only if it appears in an axiom, which is not the case.

Thus PA$_\omega$ is consistent and hence so is PA, since all theorems of PA are theorems of PA$_\omega$.

5.6 THE HEIGHT OF THIS GREAT ARGUMENT

> That to the highth of this great
> Argument
> I may assert Eternal Providence,
> And justifie the wayes of God to men.
> —Milton
> *Paradise Lost*, lines 24–26.

The consistency proof for PA depends ultimately on transfinite induction, since we assume the existence of least ordinals in order to eliminate cuts. The key ingredient in PA is induction over finite ordinals, so it may truly be said that we are using transfinite induction to justify ordinary induction. Ordinarily, this kind of "justification" would be laughed out of court, but we can defend it in this case as follows.

- We know, from Gödel's second incompleteness theorem, that induction over finite ordinals ("ω-induction") cannot prove its own consistency. Thus the strength of ω-induction can be measured, if at all, only by some ordinal *greater* than ω.

- It was shown by Gentzen (1943) that induction up to any ordinal less than ε_0 can be proved in PA.

- Therefore, if the consistency of PA can be proved by induction up to ε_0, then ε_0 is a precise measure of the strength of ω-induction. Thus the "height" of the consistency proof measures the strength of PA.

To see the details of this defence, one needs to consult a textbook on proof theory, such as Troelstra and Schwichtenberg (1996). However, we can sketch the ideas involved.

ENCODING TRANSFINITE INDUCTION IN ARITHMETIC

How can it be that ω-induction implies induction up to (some) ordinals α greater than ω? First we need to *express* transfinite induction by a sentence in the language of arithmetic; then we need to *prove* the sentence by ω-induction. Both tasks may be illustrated by the case of ω^2-induction.

We begin by setting up an ordering of the positive integers with the same order type as the ordinal ω^2. This means finding an arrangement of positive integers in the pattern of the ordinals less than ω^2:

$$1, 2, 3, \ldots \omega, \omega + 1, \omega + 2, \ldots \omega \cdot 2, \omega \cdot 2 + 1, \omega \cdot 2 + 2, \ldots \omega \cdot 3, \omega \cdot 3 + 1,$$
$$\omega \cdot 3 + 2, \ldots .$$

Perhaps the simplest such arrangement is

$$1, 3, 5, \ldots 2, 6, 10, \ldots 4, 12, 20, \ldots 8, 24, 40, \ldots .$$

We obtain this arrangement by factorizing each positive integer into a power of 2 times an odd number: $2^m(2n + 1)$. When two integers contain different powers of 2, the one with the smaller power of 2 is further to the left. And when two different integers contain the same power of 2, the one with the smaller odd factor is further to the left.

Putting this a little more formally, we denote "a is to the left of b" by the formula $a <_{\omega^2} b$, defined for positive integers a and b by

$$a <_{\omega^2} b \Leftrightarrow (\exists m)(\exists n)(\exists p)(\exists q)[a = 2^m(2n + 1) \wedge b = 2^p(2q + 1)$$
$$\wedge (m < p \vee (m = p \wedge n < q))].$$

With the help of the inductive definition of exponentiation (given in Section 2.5) we can express this ordering in the language of Peano arithmetic. Then ω^2-induction is equivalent to the statement that, *for any property P that holds for some positive integer, there is a "least" positive integer (in the sense of the ordering $a <_{\omega^2} b$) with property P.* As with ω-induction, we need to make this statement only for properties P definable in the language of PA. Thus, given a formula for P, and the formula for $a <_{\omega^2} b$ just obtained, we can express ω^2-induction as follows:

$$(\exists x)P(x) \Rightarrow (\exists y)[P(y) \wedge (\forall z)(z <_{\omega^2} y \Rightarrow \neg P(z))].$$

For each P, we can prove this statement of ω^2-induction with the help of ω-induction. To find the "least" y with property P, write each x with property P in the form $2^m(2n + 1)$. Among these numbers x, consider those with the least value of m (which exists by ω-induction). Among the latter, consider the one with the least value of n (which likewise exists by ω-induction). In this way we find the "least" x with property P (in the sense of the ordering $a <_{\omega^2} b$), thereby proving ω^2-induction by two applications of ω-induction. It is clear that the argument can be formalized in PA.

By elaborating on this argument we can express and prove α-induction in PA for many countable ordinals α, but not all. There is clearly some upper bound to the α for which such expressions exist, since there are $\aleph_1$ countable ordinals but only $\aleph_0$ formulas of PA. More interestingly, our ability to prove α-induction falls short of our ability to express it.

As mentioned above, Gentzen showed that α-induction can be proved in PA for the ordinals $\alpha < \varepsilon_0$. He had already proved the consistency of PA by ε_0-induction, and he knew that ε_0-induction can be expressed in PA, so he could conclude that ε_0-induction is expressible but *not* provable in PA—otherwise PA would be able to prove its own consistency.

Thus ε_0—the "height" of Gentzen's great consistency proof—pinpoints the *exact* position of the consistency of PA relative to the theorems of PA. If we measure the "heights" of theorems by the ordinals α for which α-induction is provable, then the height of consistency is the least upper bound of the heights of theorems. Gentzen's proof also suggests a nice way to express the consistency of PA by an arithmetic sentence; namely, by a sentence that expresses ε_0-induction. It may not be clear how to define the requisite ordering of positive integers, but in fact we already touched on a suitable method in the section on Goodstein's theorem in Chapter 2. We return to this idea in the next chapter. But first we should say a little more about ε_0.

WHY THE HEIGHT IS ε_0

We know from Section 2.1 that ε_0 is the least upper bound of the sequence

$$\omega, \quad \omega^\omega, \quad \omega^{\omega^\omega}, \quad \omega^{\omega^{\omega^\omega}}, \quad \ldots .$$

It follows that ε_0 *is the least ordinal β with the property that, if $\alpha < \beta$, then* $\omega^\alpha < \beta$. Speaking a little more loosely, ε_0 is the "least ordinal greater than ω that is closed under exponentiation."

Exponentiation is involved in the cut elimination process of Section 5.4. We start with a proof in PA, translate it into a proof in PA_ω, obtaining a tree with only finite branches but perhaps infinitely many vertices. By a careful analysis of proofs in PA, it can be checked that the corresponding proof-tree in PA_ω has ordinal $\alpha < \varepsilon_0$. Then an analysis of the cut-elimination process shows that there is a cut-free proof with ordinal $< 2^\alpha$

(see, for example, the appendix of Mendelson (1964)). Since ε_0 is closed under exponentiation, it follows that the ordinal of the cut-free proof is also $< \varepsilon_0$. Therefore, the transfinite induction we need, in order to prove the existence of a "least" cut at each step of the elimination process, is induction up to ε_0.

This is why ε_0-induction implies the consistency of PA.

5.7 ROADS TO INFINITY

The fact that transfinite induction even up to the number ε_0 is no longer derivable from the remaining number-theoretical forms of inference therefore reveals from a new angle the *incompleteness* of the number-theoretical formalism, which has already been exposed by Gödel from a different point of view.

—Gerhard Gentzen (1943),
in Gentzen (1969), pp. 307–308.

There is a remarkable parallel between the uncountability proofs in set theory and the unprovability proofs for arithmetic. In Chapter 1 we saw the first style of uncountability proof, showing that $2^{\aleph_0}$ is uncountable by diagonalization (in particular, there are uncountably many sets of positive integers). In Chapter 2 we saw a second style, showing that there are uncountably many countable ordinals. The second style of proof gives us the *smallest* uncountable cardinal $\aleph_1$, and we are as yet unable to decide whether $2^{\aleph_0} = \aleph_1$. The diagonal argument merely shows that $2^{\aleph_0}$ is uncountable; it does not determine "how far beyond countable" $2^{\aleph_0}$ is. This of course is the continuum problem.

In Chapter 3 we again used the diagonal argument to prove that any sound formal theory for arithmetic is incomplete. There are true sentences that it can express but cannot prove. Moreover, an analysis of the incompleteness proof shows that Con(PA), a sentence expressing the consistency of Peano arithmetic, is one such true but unprovable sentence. However, the diagonal argument is again a rather blunt instrument. It merely shows that Con(PA) is unprovable in PA; it does not give its "level of unprovability."

Here is how we can find the "level" of Con(PA). As we saw in Sections 5.5 and 5.6, Con(PA) is implied by ε_0-induction, and α-induction is provable in PA for each $\alpha < \varepsilon_0$. Thus ε_0 *is the least ordinal α for which α-induction is not provable in PA*. In this sense, ε_0 represents the lowest level of unprovable sentences of PA, and this is the level where ε_0-induction and Con(PA) occur.

The provability of α-induction for $\alpha < \varepsilon_0$ was shown by Gentzen (1943). In the same paper he also showed *directly*—that is, without assuming the unprovability of Con(PA), though necessarily assuming the consistency of PA—that ε_0-induction is not provable in PA.

Gentzen's direct proof that ε_0-induction is unprovable in PA beautifully completes the parallel between set theory and PA, by simulating the argument that $\aleph_1$ is the smallest uncountable cardinal. (Remember: $\aleph_1$ is the least upper bound of the countable ordinals, so if $\aleph_1$ is countable, it is less than itself.) It also adds to the evidence that ε_0 is the correct "measure" of the height of PA.

GENTZEN'S INCOMPLETENESS THEOREM

Assuming PA is consistent, ε_0-induction is not provable in PA.
On the one hand, he shows that a proof of α-induction in PA has ordinal at least α in his ordering of proofs. On the other hand, the methods of his consistency proof show that any proof in PA has ordinal less than ε_0. So, if ε_0-induction is provable in PA, then ε_0 is less than itself! This contradiction shows that ε_0-induction is not provable in PA.

To recapitulate: assuming consistency of PA, we now have two unprovable sentences of PA, arising from "miniaturization" of arguments from set theory.

- Gödel's proof that Con(PA) is unprovable in PA is a miniaturization of the argument in set theory that $2^{\aleph_0}$ is uncountable.

- Gentzen's proof that ε_0-induction is unprovable in PA is a miniaturization of the argument that $\aleph_1$ is the least uncountable cardinal.

In Gentzen's parallel between PA and set theory, ε_0 plays the role of $\aleph_1$, because ε_0 is the least ordinal that is not the height of a proof in PA.

What is *not* parallel to set theory is Gentzen's theorem that ε_0-induction implies Con(PA). This implies, as we have seen, that the sentences Con(PA) and ε_0-induction both occur at level ε_0. Thus, the diagonal construction of an unprovable sentence leads to the same level of unprovability as the ordinal construction. In set theory we still do not know the difference in level (if any) between the road to $2^{\aleph_0}$ and the road to $\aleph_1$—this is the continuum problem. But in PA all roads lead to ε_0!

5.8   HISTORICAL BACKGROUND

PEANO ARITHMETIC

For proofs in arithmetic, I used Grassmann (1861). The recent work of Dedekind (1888) was also most useful to me; in it, questions pertaining to the foundations of numbers are acutely examined.

—Giuseppe Peano
Preface to *Principles of Arithmetic* (1889),
translated by van Heijenoort (1967).

The first steps in "Peano arithmetic" were taken by Hermann Grassmann, a highly original German mathematician who worked as a high school teacher. In 1861 he published a *Lehrbuch der Arithmetik* (Textbook of Arithmetic), intended for use in schools. In it, the operations of addition and multiplication are defined, and their basic properties proved, by induction. In particular, Grassmann proves that these two operations satisfy the commutative, associative, and distributive laws:

$$a + b = b + a, \quad ab = ba,$$
$$a + (b + c) = (a + b) + c, \quad a(bc) = (ab)c,$$
$$a(b + c) = ab + ac.$$

Who says "new math" was a folly of the 1960s?! Grassmann may not have understood his pupils' needs, but there is no doubt that he understood the foundations of arithmetic better than his mathematical contemporaries. Even the greatest of them offered only naive justifications of properties such as $ab = ba$, if they bothered to justify them at all. For example, Dirichlet's cutting-edge *Vorlesungen über Zahlentheorie* (Lectures on Number Theory) of 1867 asks the reader to consider a rectangle of length a and width b in order to be convinced that $ab = ba$. Grassmann, in contrast, makes it clear that induction is *the* crucial idea in arithmetic, though without going so far as to describe induction as an axiom.

Richard Dedekind, who began his career as Dirichlet's editor, does not seem to have read Grassmann. He independently realized that induction is the crucial idea, and went further, attempting an *explanation* of the inductive property of natural numbers in terms of sets. This included the following bold attempt to *prove* that infinity exists, in his 1888 *Was sind und was sollen die Zahlen*—see Dedekind (1901), p. 64.

Theorem. There exist infinite systems.

Proof. My own realm of thoughts, i.e., the totality S of all things which can be objects of my thought, is infinite. For if s signifies an element of S, then is the thought s', that s can be an object of my thought, itself an element of S ... Hence S is infinite, which was to be proved.

I have skipped the part where Dedekind shows that the map $s \mapsto s'$ is a one-to-one correspondence between S and a proper subset of S. This part is mathematically sound and it *does* imply that S is infinite. The problem, of course, is that the concept of "things which can be objects of my thought" is not at all clear or safe from paradox. Because of this, most mathematicians regard Dedekind's "proof" as a bridge too far. But no matter, we can take a step back and take induction as an *axiom*, which is what Peano did in 1889.

Both Dedekind and Peano used the so-called *second order* induction axiom, in which the property $\mathcal{P}$ is an arbitrary property of natural numbers, not necessarily one definable in the formal language of arithmetic. The second order axiom is convenient because it can be written as a single formula,

$$(\forall \mathcal{P}) \left[[\mathcal{P}(1) \wedge (\forall n)(\mathcal{P}(n) \Rightarrow \mathcal{P}(n+1))] \Rightarrow (\forall n)\mathcal{P}(n) \right].$$

The problem with second order induction, not realized at first, is that the underlying second order logic is not as well-behaved as predicate logic. Second order logic allows $\forall$ and $\exists$ to apply to properties as well as individuals, and this rules out completeness and compactness theorems, among other things.

The idea (in Section 5.4) of cutting back to *first order* axioms, in which $\mathcal{P}$ is expressible by a formula and one needs a whole infinite schema of induction axioms, emerged under the influence of predicate logic in the 1920s. Predicate logic allows the quantifiers $\forall$ and $\exists$ to be applied only to individuals, not properties, and this brings many advantages. However, its advantages did not become apparent until Gödel proved the completeness and compactness theorems for predicate logic in 1929.

The advantages were reinforced by his incompleteness theorems of 1930, which implies incompleteness of second order logic, and by Gentzen's consistency proof for first order arithmetic, Gentzen (1936). Gentzen actually hoped to prove consistency of second order arithmetic (which was what Hilbert originally wanted), but did not succeed. If he had, perhaps we would still define "Peano arithmetic" as Peano did!

However, this is no longer likely to happen. There is much less chance of a Gentzen-style consistency proof for second order arithmetic, say, using induction up to some countable ordinal. In any case, second order arithmetic is *much more* than number theory—it contains a great deal of analysis. Being able to talk about arbitrary sets of natural numbers is equivalent to being able to talk about arbitrary real numbers. Many classical theorems of analysis, such as the Bolzano-Weierstrass and Heine-Borel theorems, can be proved in quite small fragments of second order arithmetic. For a thorough study of second order arithmetic and its fragments, with special attention to analysis, see Simpson (1999).

ORDINALS AND ARITHMETIC

> Now it also becomes clear why it is the rule of transfinite induction that is
> needed as the crucial rule for the consistency proof ... Proof number 1 is
> after all trivially correct; and once the correctness of all proofs preceding a
> particular proof has been established, the proof in question is also correct
> precisely because the ordering [of proofs] was chosen in such a way that the
> correctness of a proof depends on the correctness of certain earlier proofs.
>
> —Gerhard Gentzen
> *The Concept of Infinity in Mathematics,*
> in Gentzen (1969), p. 232.

The work of Borel and Hardy from around 1900, relating ordinals to the
growth rate of integer functions, ultimately proved to be an excellent
idea. But it did not yield much fruit at first. As mentioned in Section 2.3,
Hardy erred in thinking that all countable ordinals can be generated in
some effective way. His confusion was not completely dispelled by Zermelo's recognition of the role of the axiom of choice in well-ordering in
1904. More than 20 years later, Hilbert's attempt to prove the continuum
hypothesis in his *On the Infinite* was marred by confusion about countable
ordinals.

As far as I know, Gentzen's introduction of ordinals in his 1936 consistency proof for PA was a completely new idea. He set off in this direction
only after making an entirely different attempt to prove the consistency
of PA by reducing it to an arithmetic based on the "intuitionist" principles of Brouwer. In effect, he proved in 1933 that if intuitionist arithmetic
is consistent, then so is PA. This was a surprise, because intuitionist principles were supposed to be more reliable than classical mathematics. Yet,
somehow they are just as strong as PA, whose consistency cannot be
proved in PA. Does this mean we should have more faith in PA, or less
faith in intuitionist arithmetic?

Gentzen's proof was never published, because Gödel had the same
idea independently, and published it in 1933. When Gentzen learned of
Gödel's priority, he withdrew his own paper on intuitionist arithmetic
and made the fateful turn towards cut elimination. He succeeded in
eliminating cuts from predicate logic in 1934, and proved consistency of
PA in 1936, thanks to his brilliant use of countable ordinals. He was able
to avoid working with *all* the countable ordinals because PA calls only for
ordinals up to ε_0. As a byproduct of his proof, he discovered a new type
of unprovable sentence of PA, namely, ε_0-induction.

While ε_0-induction is an interesting unprovable sentence of PA, it
is open to the same aesthetic objection as Gödel's unprovable sentence
Con(PA): such sentences do not arise naturally in the study of arithmetic. The theorem of Goodstein (1944) comes closer to the goal of
a "natural" unprovable sentence, since it involves only the concepts of

addition, multiplication, and exponentiation on positive integers. As we saw in Section 2.7, Goodstein realized that his theorem is a consequence of ε_0-induction. In the next chapter we give a natural generalization of Goodstein's theorem that *implies* ε_0-induction, hence the generalization is certainly not provable in PA (if PA is consistent).

The original Goodstein theorem is not provable in PA either, but this fact is more difficult to establish. It took three further decades of research on the relation between ordinals and PA before Goodstein's theorem was properly understood. The first step was taken by the Austrian-born (but English-educated) mathematician Georg Kreisel in 1952, who discovered a relationship between *provably computable* functions and *ordinal recursive* functions.

A computable function f is said to be *provably* computable if PA proves that the computation of $f(m)$ halts for each natural number m. It is possible to concoct a computable function that is not provably computable by listing all the theorems of the form "the computation of $f(m)$ halts for all m" and making a diagonal construction. However, the diagonal argument does not tell us what, if anything, makes the provably computable functions special. For that we need Gentzen's analysis of proofs in PA.

Like any provable sentence of PA, "the computation of $f(m)$ halts for all m" necessarily has some ordinal $\alpha < \varepsilon_0$. It is to be expected that α somehow bounds how "complex" f can be, but how do we measure complexity? One way is to define computable functions by α-induction for countable ordinals α. Just as we can express α-induction in PA for $\alpha < \varepsilon_0$, we can define functions in PA by α-induction for $\alpha < \varepsilon_0$. Assuming that the non-inductive part of the definition is in some sense simple, such functions f are computable, in which case they are called *ordinal recursive*. Kreisel's discovery was that the provably computable functions are precisely the ordinal recursive functions for ordinals $\alpha < \varepsilon_0$.

Another way to measure complexity of computable functions is by rate of growth. This was discovered by Stanley Wainer (1970), elegantly connecting provability with the old growth-rate ideas of Borel and Hardy. Wainer showed that the provably computable functions are precisely the computable functions whose growth rate is bounded by a function H_α in the so-called *Hardy hierarchy*, for some $\alpha < \varepsilon_0$. (For more on the Hardy hierarchy, see Section 6.5.) Thus ε_0 again plays the role of least upper bound to the complexity of proofs in Peano arithmetic.

It follows from Wainer's theorem that, *if f is a computable function that grows faster than any H_α for $\alpha < \varepsilon_0$, then f is not provably computable in PA*. This allows us to find many new examples of sentences not provable in PA, as we will see in the next chapter.

Chapter 6

Natural Unprovable Sentences

Preview

In this chapter we return to Goodstein's theorem, the result proved in Chapter 2 with the help of ε_0-induction. We find a natural generalization of Goodstein's theorem that is not only a consequence of ε_0-induction, but also *equivalent* to it. It follows that the generalized Goodstein theorem implies the consistency of PA, and hence it is not provable in PA. This gives us our first example of a natural theorem not provable in PA.

The ordinary Goodstein theorem is itself not provable in PA, but showing this requires a more delicate analysis of ordinals and proofs, involving fast-growing functions associated with countable ordinals. We outline how this is done for Goodstein's theorem, and describe some other theorems whose unprovability in PA is similarly tied to fast-growing functions.

These other unprovable theorems lie in popular areas of combinatorics—Ramsey theory and graph theory. In both areas we exhibit natural problems involving fast-growing functions. The Ramsey theory example has a growth rate similar to that of the Goodstein function, so the corresponding unprovable theorem—the *Paris-Harrington theorem*—lies at a level similar to that of Goodstein's theorem.

In graph theory there are two remarkable theorems, *Kruskal's theorem* and the *graph minor theorem*, lying at an even higher level. Thus the latter theorems are provable only from assumptions about infinity much stronger than ε_0-induction.

6.1   A GENERALIZED GOODSTEIN THEOREM

Recall the *Goodstein process* introduced in Section 2.7. It starts with a positive integer m_1 written in base 2 Cantor normal form. For example, we might start with

$$m_1 = 27 = 2^4 + 2^3 + 2 + 1 = 2^{2^2} + 2^{2+1} + 2 + 1.$$

Next, replace each 2 by a 3, obtaining (in this case)

$$n_1 = 3^{3^3} + 3^{3+1} + 3 + 1,$$

and subtract 1. The resulting number is written in base 3 Cantor normal form, in this case

$$m_2 = 3^{3^3} + 3^{3+1} + 3.$$

Then the process continues with the new base 3, replacing it by 4 and again subtracting 1. In a nutshell, the Goodstein process produces a sequence of numbers in Cantor normal form with respective bases 2, 3, 4, 5, . . . , by alternately replacing each k in a base k number by $k + 1$, and subtracting 1. *Goodstein's theorem* states that the process terminates for any initial number m_1.

The proof, as we saw in Section 2.7, is easy once one spots the countable ordinals lurking in the shadows of the Goodstein process. We can associate each base k Cantor normal form number m_{k-1} with a *Cantor normal form ordinal* α_{k-1} by replacing each k in the base k expression for m_{k-1} by ω. In the present case, for example,

$$m_1 = 2^{2^2} + 2^{2+1} + 2 + 1, \quad \text{so} \quad \alpha_1 = \omega^{\omega^\omega} + \omega^{\omega+1} + \omega + 1.$$

It follows that *the ordinal remains the same when we replace k by $k + 1$*, and *the ordinal decreases when we subtract 1*. Thus, continuing the present example, we have

$$m_2 = 3^{3^3} + 3^{3+1} + 3, \quad \text{and} \quad \alpha_2 = \omega^{\omega^\omega} + \omega^{\omega+1} + \omega.$$

Goodstein's theorem follows because *any decreasing sequence of ordinals is finite*. In fact, since the Cantor normal form ordinals used in the proof are all less than ε_0, we need assume the decreasing sequence property only for the ordinals less than ε_0. In other words, Goodstein's theorem follows by what may be called *transfinite induction up to ε_0*.

As soon as one sees the connection with decreasing sequences of ordinals, it becomes clear that many variations of the Goodstein process terminate for exactly the same reason. Instead of replacing k with $k + 1$

one can replace k by *any* integer $k' \geq k$ and the ordinal does not change, so the process still must terminate if these base increases alternate with subtractions of 1. Let us call a process that alternates between replacing each k in the base k Cantor normal form of an integer by any $k' \geq k$ (thus obtaining an integer in base k' normal form) with subtraction of 1 a *general Goodstein process*. Then we have:

GENERALIZED GOODSTEIN THEOREM. *A general Goodstein process, applied to any positive integer, terminates in a finite number of steps.*

Proof: By transfinite induction up to ε_0.   □

6.2 COUNTABLE ORDINALS VIA NATURAL NUMBERS

The gist of the previous section is that ε_0-induction implies the generalized Goodstein theorem. In fact, we can also prove that the generalized Goodstein theorem implies ε_0-induction, so generalized Goodstein and ε_0-induction are "equivalent" in some sense. The exact meaning of "equivalent" depends on how much we need to assume in proving the two implications—a question we defer for the moment. First, we need to see that generalized Goodstein implies ε_0-induction. After we have some kind of proof, we can look more closely at our assumptions.

To find a route from the generalized Goodstein theorem to ε_0-induction, we begin by looking at a table of Cantor normal forms of all the positive integers. Table 6.1 is the beginning of such a table, whose leftmost column lists (in the upward direction) all the positive integers, and whose second, third, fourth, ... columns contain the Cantor normal forms of each positive integer in bases 2, 3, 4, ... respectively.

Each entry in column k is a base k Cantor normal form, that is, an instance of a classical Cantor normal form, obtained by replacing each symbol ω by k. For example, the base 2 Cantor normal form $2^2 + 1$ of 5 is an instance of the classical Cantor normal form $\omega^\omega + 1$, and the base 3 Cantor normal form $3^2 + 1$ of 10 is an instance of the classical Cantor normal form $\omega^2 + 1$. Conversely, each classical Cantor normal form φ has instances in the table. They occur for precisely the bases k greater than all the integers occurring in the form φ. For example, there is an instance of the Cantor normal form

$$\varphi = \omega^2 + 1$$

in all columns $k > 2$, namely the number $k^2 + 1$.

Table 6.2 shows the classical Cantor normal forms associated with the entries in Table 6.1.

$\vdots$	$\vdots$	$\vdots$	$\vdots$	$\vdots$	$\vdots$
16	2^{2^2}	$3^2+3\cdot2+1$	4^2	$5\cdot3+1$	$6\cdot2+4$
15	$2^{2+1}+2^2+2+1$	$3^2+3\cdot2$	$4\cdot3+3$	$5\cdot3$	$6\cdot2+3$
14	$2^{2+1}+2^2+2$	3^2+3+2	$4\cdot3+2$	$5\cdot2+4$	$6\cdot2+2$
13	$2^{2+1}+2^2+1$	3^2+3+1	$4\cdot3+1$	$5\cdot2+3$	$6\cdot2+1$
12	$2^{2+1}+2^2$	3^2+3	$4\cdot3$	$5\cdot2+2$	$6\cdot2$
11	$2^{2+1}+2+1$	3^2+2	$4\cdot2+3$	$5\cdot2+1$	$6+5$
10	$2^{2+1}+2$	3^2+1	$4\cdot2+2$	$5\cdot2$	$6+4$
9	$2^{2+1}+1$	3^2	$4\cdot2+1$	$5+4$	$6+3$
8	2^{2+1}	$3\cdot2+2$	$4\cdot2$	$5+3$	$6+2$
7	2^2+2+1	$3\cdot2+1$	$4+3$	$5+2$	$6+1$
6	2^2+2	$3\cdot2$	$4+2$	$5+1$	6
5	2^2+1	$3+2$	$4+1$	5	5
4	2^2	$3+1$	4	4	4
3	$2+1$	3	3	3	3
2	2	2	2	2	2
1	1	1	1	1	1
	base 2	base 3	base 4	base 5	base 6 $\quad\cdots$

Table 6.1. Base 2, 3, 4, … Cantor normal forms of the positive integers.

Now forget for the moment that classical Cantor forms represent ordinals. Let us see what it takes to *prove* that these objects are well ordered.

First we define the ordering of Cantor normal forms. This can be done by ordinary induction on what may be called the *height* of the form—the number of levels of exponents. The forms of height 0 are simply the positive integers 1, 2, 3, 4, …, and these are ordered in the usual way. Then if

$$\varphi = \omega^{\alpha_1}\cdot m_1 + \omega^{\alpha_2}\cdot m_2 + \cdots + \omega^{\alpha_i}\cdot m_i$$

is a form of height $h+1$, this means that the maximum height of $\alpha_1, \alpha_2, \ldots, \alpha_i$ is h. We may therefore assume, by induction, that the order of these exponents is already defined, and we may take it that

$$\alpha_1 \succ \alpha_2 \succ \cdots \succ \alpha_i,$$

where $\succ$ denotes the ordering of classical Cantor normal forms. Given another form of height $h+1$,

$$\psi = \omega^{\beta_1}\cdot n_1 + \omega^{\beta_2}\cdot n_2 + \cdots + \omega^{\beta_j}\cdot n_j,$$

⋮	⋮	⋮	⋮	⋮	⋮
16	ω^{ω^ω}	$\omega^2 + \omega \cdot 2 + 1$	ω^2	$\omega \cdot 3 + 1$	$\omega \cdot 2 + 4$
15	$\omega^{\omega+1} + \omega^\omega + \omega + 1$	$\omega^2 + \omega \cdot 2$	$\omega \cdot 3 + 3$	$\omega \cdot 3$	$\omega \cdot 2 + 3$
14	$\omega^{\omega+1} + \omega^\omega + \omega$	$\omega^2 + \omega + 2$	$\omega \cdot 3 + 2$	$\omega \cdot 2 + 4$	$\omega \cdot 2 + 2$
13	$\omega^{\omega+1} + \omega^\omega + 1$	$\omega^2 + \omega + 1$	$\omega \cdot 3 + 1$	$\omega \cdot 2 + 3$	$\omega \cdot 2 + 1$
12	$\omega^{\omega+1} + \omega^\omega$	$\omega^2 + \omega$	$\omega \cdot 3$	$\omega \cdot 2 + 2$	$\omega \cdot 2$
11	$\omega^{\omega+1} + \omega + 1$	$\omega^2 + 2$	$\omega \cdot 2 + 3$	$\omega \cdot 2 + 1$	$\omega + 5$
10	$\omega^{\omega+1} + \omega$	$\omega^2 + 1$	$\omega \cdot 2 + 2$	$\omega \cdot 2$	$\omega + 4$
9	$\omega^{\omega+1} + 1$	ω^2	$\omega \cdot 2 + 1$	$\omega + 4$	$\omega + 3$
8	$\omega^{\omega+1}$	$\omega \cdot 2 + 2$	$\omega \cdot 2$	$\omega + 3$	$\omega + 2$
7	$\omega^\omega + \omega + 1$	$\omega \cdot 2 + 1$	$\omega + 3$	$\omega + 2$	$\omega + 1$
6	$\omega^\omega + \omega$	$\omega \cdot 2$	$\omega + 2$	$\omega + 1$	ω
5	$\omega^\omega + 1$	$\omega + 2$	$\omega + 1$	ω	5
4	ω^ω	$\omega + 1$	ω	4	4
3	$\omega + 1$	ω	3	3	3
2	ω	2	2	2	2
1	1	1	1	1	1
	base 2	base 3	base 4	base 5	base 6 ⋯

Table 6.2. Classical Cantor normal forms for the entries in Table 6.1.

we may likewise assume that $\beta_1 \succ \beta_2 \succ \cdots \succ \beta_j$, and that we know the order of all the exponents α and β. Then $\varphi \succ \psi$ if and only if $\alpha_k \succ \beta_k$ at the first exponent where φ and ψ differ, or (if all exponents are the same) that $m_j \succ n_j$ at the first coefficient where they differ (which is the same as $m_j > n_j$, since m_j and n_j are integers).

It is clear from the definition of this ordering that the Cantor normal forms are *linearly*, or *totally*, ordered. That is, if φ and ψ are two forms then either $\varphi \succ \psi$ or $\psi \succ \varphi$. In fact, the definition of the ordering $\succ$ and the proof that it is linear can be carried out in Peano Arithmetic (the theory PA described in Section 5.3). The Cantor normal forms can be encoded by natural numbers, and the linear order properties can be proved with the help of ordinary induction and properties of addition, multiplication, and exponentiation provable in PA.

It is a different matter to prove that Cantor normal forms are *well* ordered. This is precisely the well-ordering property of ε_0, so it is *not* provable in PA for the reasons we saw in Chapters 3 and 5. The well-ordering of ε_0 implies the consistency of PA (Chapter 5), which is not provable in PA by Gödel's second incompleteness theorem (Chapter 3).

Nevertheless, it is possible to deduce the well-ordering of Cantor normal forms from certain natural statements about natural numbers. One of these is the generalized Goodstein theorem, as we will see in the next section.

6.3   FROM GENERALIZED GOODSTEIN TO WELL-ORDERING

The Cantor normal forms of base 2, 3, 4, ... obviously have a close relationship with the classical Cantor normal forms, and a general Goodstein process is thereby related to a process of descent through classical Cantor normal forms. When these relationships are clarified, a deduction of ε_0 well-ordering from the generalized Goodstein theorem comes into view.

In fact, one can catch a glimpse of the relationship by staring at Tables 6.1 and 6.2. The classical Cantor normal forms are "graphs" (functions of the base) visible as points on curves that can be drawn through Table 6.2. For example: the values ω lie on a line of slope 1 starting in the "base 2" column; the values $\omega + 1$ lie on a parallel line, also starting in the "base 2" column; the values $\omega \cdot 2$ lie along a line of slope 2 starting in the "base 3" column; the values ω^2 lie along a parabola starting in the "base 3" column; and so on. It appears, from Tables 6.1 and 6.2, that the graphs of the classical Cantor normal forms are stacked up in their correct order, and that this ordering arises from the ordering of numbers within the columns of Table 6.1. Figure 6.1 shows the graphs for some of the normal forms, and their ordering.

The main point to be clarified is the relationship between the ordering $\succ$ of classical Cantor normal forms and the ordering $>$ of natural numbers. To describe this relationship we denote classical normal forms by Greek letters such as $\alpha, \beta, \ldots$, and let $\alpha(k)$ denote the base k Cantor normal form that results from α by replacing ω by k. For example, if

$$\alpha = \omega^\omega + \omega^2 + 3$$

then

$$\alpha(4) = 4^4 + 4^2 + 3, \quad \alpha(5) = 5^5 + 5^2 + 3, \quad \text{and so on.}$$

To avoid getting expressions that are not base k Cantor normal forms, we define $\alpha(k)$ only for k greater than the positive integers occurring in α. Then we have:

COMPATIBILITY OF $\succ$ AND $>$. *If α and β are any classical Cantor normal forms and k is a positive integer greater than the integers occurring in α and β, then $\alpha \succ \beta$ if and only if $\alpha(k) > \beta(k)$.*

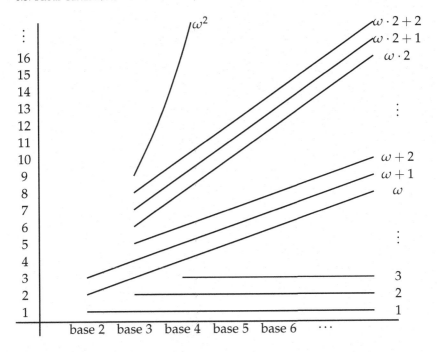

Figure 6.1. Cantor normal forms as graphs.

This is proved by an ordinary induction on height, along similar lines to the definition of $\succ$ in the previous section. It follows from the compatibility of $\succ$ and $>$ that, *if $\alpha \succ \beta$, each occurrence of $\alpha(k)$ in Table 6.1 lies above the corresponding occurrence of $\beta(k)$* (if it exists). Therefore, if $\alpha(k)$ is a number in column k of Table 6.1 representing α, and $\alpha \succ \beta$, then we can travel from $\alpha(k)$ to a number $\beta(l)$ representing β as follows:

- Replace each k in the base k form $\alpha(k)$ by l, where l is greater than k and all the integers in the Cantor normal form β (so that $\beta(l)$ exists).

- Subtract 1 from the resulting form $\alpha(l)$ (keeping the base fixed at l) $\alpha(l) - \beta(l)$ times, thus obtaining $\beta(l)$.

In the more visual terms of Figure 6.1, we descend from the graph of α to the graph of $\beta < \alpha$ by first traveling to the right on the graph of α, from base k to a base l for which the graph of β is defined, then dropping vertically down the base l column. For example, if we are on the graph of ω at base 2 and want to descend to the graph of 3, we must first increase the base to 4. Then subtract 1 to drop from the graph of ω to the graph of 3 in the base 4 column.

Making a series of such steps, we obtain the theorem:

GENERALIZED GOODSTEIN IMPLIES ε_0-INDUCTION. *Any descending sequence of classical Cantor normal forms determines a generalized Goodstein process. (So any decreasing sequence of such forms is finite).*

Proof: Given a descending sequence $\alpha \succ \beta \succ \gamma \succ \cdots$ of classical Cantor normal forms, we find positive integers $k < l < m < \cdots$ so that $\alpha(k), \beta(l), \gamma(m), \ldots$ are all defined. Namely, find

$$k > \text{the integers in the form } \alpha,$$
$$l > k \text{ and the integers in the form } \beta,$$
$$m > k, l \text{ and the integers in the form } \gamma,$$
$$\vdots$$

Then we have the following generalized Goodstein process:

- Replace k in $\alpha(k)$ by l, obtaining $\alpha(l)$.

- Subtract 1 (keeping the base l fixed) $\alpha(l) - \beta(l)$ times, obtaining $\beta(l)$.

- Replace l in $\beta(l)$ by m, obtaining $\beta(m)$.

- Subtract 1 (keeping the base m fixed) $\beta(m) - \gamma(m)$ times, obtaining $\gamma(m)$.

- And so on.

By the generalized Goodstein theorem, this process terminates. Therefore, the sequence $\alpha \succ \beta \succ \gamma \succ \cdots$ of Cantor normal forms must be finite.

It follows, finally, that any set of Cantor normal forms has a least member, and hence ε_0 is well-ordered. Because, in any infinite set of forms without least member we can construct an infinite descending sequence, by repeatedly choosing smaller members. □

6.4 GENERALIZED AND ORDINARY GOODSTEIN

The generalized Goodstein theorem, discussed in the last section, is not provable in PA. This can be seen from the fact that it implies ε_0-induction, which implies the consistency of PA by Gentzen's consistency proof, and we know that consistency of PA is not provable in PA by Gödel's second incompleteness theorem. However, the generalized Goodstein theorem is not even *expressible* in the language of PA, because it involves quantification over arbitrary sequences of positive integers, and PA has variables only for individual integers, not for sequences of them. Perhaps it would be better to say that the generalized Goodstein theorem is *stronger* than PA.

Alternatively, we can cut down the generalized Goodstein theorem to one expressible in the language of PA by replacing the arbitrary sequence of bases $k \leq k' \leq k'' \leq \cdots$ in the statement of the generalized Goodstein theorem by an arbitrary *formula* of PA defining a nondecreasing sequence of positive integers. Thus the single sentence expressing the generalized Goodstein theorem is replaced by a *Goodstein axiom schema* that includes all instances of the generalized Goodstein theorem expressible in PA.

It follows, by the argument in the previous section, that the Goodstein axiom schema implies all instances of ε_0-induction expressible in PA. The latter instances suffice to prove consistency of PA by the argument of Chapter 5. Moreover, the implications can also be proved in PA. Thus *it can be proved in PA that the Goodstein axiom schema implies the consistency of PA*. This shows, by Gödel's second theorem again, that the Goodstein axiom schema is *not* provable in PA.

We have a harder row to hoe if we start by assuming only the ordinary Goodstein theorem, where the ascending sequence of bases is 2, 3, 4, 5, ... In this case we can prove finiteness only for a restricted class of descending sequences $\alpha \succ \beta \succ \gamma \succ \cdots$ of Cantor normal forms. This restricted property of Cantor normal forms happens to be unprovable in PA, but not for well-known reasons like Gödel's second theorem, so it is desirable to find another approach to unprovability in PA. One very useful approach, which we outline in the next section, is via the study of provably computable functions. This approach allows us to show unprovability in PA of several interesting sentences, not just Goodstein's theorem. Provably computable functions are also interesting for their own sake, because they throw new light on the relationship between ordinals and growth-rate.

6.5 Provably Computable Functions

The ordinary Goodstein theorem says that we have a well-defined function

$$G(n) = \text{number of steps before the Goodstein process on } n \text{ halts.}$$

The function $G(n)$ is obviously computable, simply by letting the Goodstein process run and counting the steps. But we cannot prove in PA that

$G(n)$ exists for each positive integer n *unless* we can prove that the Goodstein process always halts. If the ordinary Goodstein theorem is provable in PA then $G(n)$ is what is known as a provably computable function.

These functions were already discussed in Section 5.8. But, to recapitulate, we call a function f *provably computable* if f is computable and PA proves that the computation of $f(m)$ halts for each m.

Provably computable functions turn out to be interesting and useful because their growth rates can be measured on a "scale" of functions labelled by ordinals $< \varepsilon_0$. Any provably computable function has growth rate less than some member of the scale; therefore, any computable function that grows so fast as to be "off the scale" is not provably computable. It follows that *any sentence implying the existence of a computable function off the scale is not provable in PA.*

We outline a description of the scale, and the reasons for its existence, in the subsection below. It turns out that the Goodstein function $G(n)$ is off the scale, *hence Goodstein's theorem is not provable in PA.*

The Scale of Provably Computable Functions

If the function f is provably computable in PA, then the corresponding proof has some ordinal $\alpha < \varepsilon_0$, by Gentzen's consistency proof for PA. Since f is computable, and PA has the resources to prove the results of all computations, PA proves all the theorems of the form

$$f(0) = a, \quad f(1) = b, \quad f(2) = c, \quad \dots,$$

where $a, b, c, \dots$ are the values of f. It follows, by Gentzen's ordering of proofs, that the theorems $f(0) = a, f(1) = b, f(2) = c, \dots$ all have ordinals bounded by $\alpha < \varepsilon_0$. Then a careful analysis of the way that proofs simulate computations shows that the sequence $a, b, c, \dots$ of values of f cannot grow very fast.

In fact, f grows more slowly than some function in the following sequence $H_0, H_1, H_2, \dots, H_\alpha, \dots$ of increasingly fast-growing functions, where $\alpha < \varepsilon_0$. This sequence of functions is also known as the *Hardy hierarchy* because its members coincide with functions used by Hardy (1904) in his attempt to describe a sequence of $\aleph_1$ real numbers "explicitly." (As mentioned in Section 2.3, Hardy mistakenly thought that H_α could be explicitly defined for all countable α, overlooking the need for the axiom of choice.)

The bottom function of the Hardy hierarchy is the identity function,

$$H_0(n) = n.$$

If H_α is defined for some ordinal α then we define $H_{\alpha+1}$ by

$$H_{\alpha+1}(n) = H_\alpha(n + 1).$$

Finally, for each $\lambda < \varepsilon_0$ we choose a sequence $\lambda_1 < \lambda_2 < \lambda_3 < \cdots$ with limit λ (based on the Cantor normal form) and let

$$H_\lambda(n) = H_{\lambda_n}(n).$$

Thus H_λ *diagonalizes* the sequence of functions $H_{\lambda_1}, H_{\lambda_2}, H_{\lambda_3}, \ldots$.
The functions in the Hardy hierarchy grow ever so slowly at first:

$$H_0(n) = n,$$
$$H_1(n) = n + 1,$$
$$H_2(n) = n + 2,$$
$$\vdots$$

and hence (taking the obvious sequences with limits $\omega, \omega \cdot 2, \omega \cdot 3, \ldots, \omega^2$)

$$H_\omega(n) = n + n = n \cdot 2,$$
$$H_{\omega+1}(n) = (n+1) \cdot 2,$$
$$\vdots$$
$$H_{\omega \cdot 2}(n) = (n+n) \cdot 2 = n \cdot 2^2,$$
$$H_{\omega \cdot 3}(n) = n \cdot 2^3,$$
$$\vdots$$
$$H_{\omega^2}(n) = n \cdot 2^n.$$

But their growth accelerates with the advent of the exponential function. Arbitrarily high towers of exponentials quickly form, then diagonalization leads to functions growing faster than any encountered in everyday mathematics. Nevertheless, all the functions H_α, for $\alpha < \varepsilon_0$, are provably computable.

Conversely, for any provably computable function f, Gentzen's ordering of proofs shows that f grows more slowly than H_α for some $\alpha < \varepsilon_0$. Thus *the Hardy functions H_α form a scale that "measures" all the provably computable functions.* How remarkable that Hardy's idea was so wrong for its original purpose, yet so right for a purpose he could not imagine!

Why the Goodstein Function Is off the Scale

Goodstein's theorem concerns the *ordinary Goodstein process* that starts with a number M in base 2 Cantor normal form, then alternately increases the base by 1 and subtracts 1. It will now be useful to consider the *arbitrary base Goodstein process* that starts with a number N in base n

Cantor normal form before alternately increasing the base by 1 and subtracting 1. It is clear that the Goodstein process starting from N and n is the tail end of an ordinary Goodstein process. Simply *run the Goodstein process backwards* from number N and base n, alternately adding 1 to the number and subtracting 1 from the base, until one reaches a number M in base 2 Cantor normal form.

It follows that the ordinary Goodstein theorem is equivalent (in PA) to the *arbitrary base Goodstein theorem*, stating that the arbitrary base Goodstein process reaches zero for each positive integer N and each base $n \geq 2$.

We now show that the arbitrary base Goodstein theorem is not provable in PA because it involves a function that is not provably computable. In fact, the function in question overtakes every Hardy function H_α for $\alpha < \varepsilon_0$. This is due to a common feature of the Goodstein process and the construction of the Hardy functions H_α for $\alpha < \varepsilon_0$. Both are based on the Cantor normal form. Also based on the Cantor normal form is the *n-predecessor* function $P_n(\alpha)$ for ordinals $\alpha < \varepsilon_0$, which reflects subtraction of 1 from the base n Cantor normal form integer corresponding to α.

For the Goodstein process that starts with the base n Cantor normal form corresponding to α, then alternately increases the base by 1 and subtracts 1, the reflected ordinal predecessor functions are $P_{n+1}, P_{n+2}, \ldots$. It follows that *the arbitrary base Goodstein theorem is equivalent to the sentence*

$$(\forall n)(\exists m)[P_m P_{m-1} \cdots P_{n+2} P_{n+1}(\alpha) = 0], \quad \text{(for all } \alpha < \varepsilon_0),$$

which expresses termination of the process at ordinal zero. (One might expect base increase operations to be involved as well, but happily they cancel out of the final equation.)

Associated with this sentence we have the computable function

$$f_\alpha(n+1) = \text{least } m \text{ such that } [P_m P_{m-1} \cdots P_{n+2} P_{n+1}(\alpha) = 0].$$

It turns out (by an ingenious suite of interlocking induction arguments, due to Cichon (1983)) that $f_\alpha(n+1)$ is none other than $H_\alpha(n+1) - 1$. This is where the Hardy functions get into the act, as promised.

Of course, the Hardy function $H_\alpha(n+1)$ is provably computable for any fixed $\alpha < \varepsilon_0$. But the arbitrary base Goodstein theorem, if provable in PA, implies that $H_\alpha(n+1)$ is also provably computable *as a function of* α. If this is so, then specializing to

$$\alpha = \omega, \quad \omega^\omega, \quad \omega^{\omega^\omega}, \quad \ldots$$

shows that the diagonal function $H_{\varepsilon_0}(n)$ of

$$H_\omega(n), \quad H_{\omega^\omega}(n), \quad H_{\omega^{\omega^\omega}}(n), \quad \ldots$$

is a provably computable function of n. It isn't—because H_{ε_0} grows faster than any function in the Hardy hierarchy. Thus the arbitrary base Goodstein theorem, and hence the ordinary Goodstein theorem, is not provable in PA.

6.6 COMPLETE DISORDER IS IMPOSSIBLE

> Whereas the entropy theorems of probability theory and mathematical physics imply that, in a large universe, disorder is probable, certain combinatorial theorems imply that complete disorder is impossible.
> —Theodore Motzkin (1967), p. 244.

The subject of *Ramsey theory* can be (and often is) introduced by a simple fact: *in any group of six people there are either three mutual acquaintances or three mutual strangers*. This fact can be presented and proved graphically as follows. Represent the people by dots (called *vertices*), and connect any two dots by a line (called an *edge*)—a black line if the corresponding people know each other, a gray line if they do not. The result is what we call the *complete graph on six vertices*, K_6, with its edges colored in two colors (gray and black).

Then the fact we claim is that *the 2-colored K_6 contains a monochromatic triangle*, that is, either a gray triangle or a black triangle.

To see why this is true, first observe that each vertex in K_6 lies at the end of five edges (to the other five vertices). Therefore, in the 2-colored K_6, at least three of these edges have the same color. Figure 6.2 shows a typical case, where three edges from the same vertex are black.

The only way that our K_6 can now avoid containing a black triangle is for each edge joining the ends of a pair of these black edges to be gray. But in this case we get a gray triangle, as Figure 6.3 shows.

Thus, monochromatic triangles are unavoidable.

In a small way, this example illustrates Motzkin's saying that *complete disorder is impossible*. "Order" here is monochromatic triangles, and the proof shows that 2-colored complete graphs with at least six vertices cannot avoid it. (Such order *is* avoidable with five vertices. Can you see how?) In graph theory language, a triangle is a K_3 (complete graph on

Figure 6.2. Monochrome edges from the same vertex.

Figure 6.3. Avoiding black triangles.

three vertices), so we have: *for any $n \geq 6$, any 2-colored K_n contains a monochromatic K_3*.

When the example is expressed like this, generalizations are irresistible. We define the *complete graph K_n* to be the graph with n vertices and an edge between any two of them. Then we expect that *for any m there is an N such that, for any $n \geq N$, a 2-colored K_n contains a monochromatic K_m.* That is, in any sufficiently large group of people, there are either m mutual acquaintances, or m mutual strangers. In particular, for the case $m = 4$ it is known that any group of 18 or more people contains either four mutual acquaintances or four mutual strangers.

Next, abandoning the acquaintance/stranger story, we consider coloring the edges of the complete graph with k colors, in which case we claim that: *for any k and m there is an N such that, for any $n \geq N$, a k-colored K_n contains a monochromatic K_m.*

Finally, instead of assigning a color to each edge, which corresponds to a pair of points, we can consider assigning a color to each i-element set of points. This at last brings us to the English mathematician Frank Ramsey, who in 1930 proved the mother of all the little Ramsey theorems above:

FINITE RAMSEY THEOREM. *For any i, k, m there is an N such that, if the i-element subsets of a set with N or more elements are colored in k colors, then there is a subset of size m whose i-element subsets all have the same color.*

Ramsey (1930) also proved an infinite Ramsey theorem:

INFINITE RAMSEY THEOREM. *For any i, k, and any k-coloring of the i-element subsets of a countably infinite set, there is an infinite subset whose i-element subsets all have the same color.*

There is a remarkable connection between the finite and infinite versions: *the finite Ramsey theorem follows from the infinite Ramsey theorem by the König infinity lemma.* This connection was not noticed by Ramsey, who gave separate proofs of the two versions. The connection between finite and infinite is a "compactness" phenomenon like that observed for propositional and predicate logic in Section 4.7. Somewhat surprisingly, the connection is in the direction from infinite to finite for Ramsey theory. König originally envisaged that his lemma would be used in the

From infinite Ramsey to finite Ramsey

To save words, let us call a k-coloring of the i-element subsets of a set S simply a "coloring of S."

Suppose that the finite Ramsey theorem is false. Then there is an m such that, for any n, there is a coloring of $\{1, 2, \ldots, n\}$ with no monochromatic subset of size m. We take all such colorings to be vertices of a tree T, with a single "root" vertex (corresponding to the empty set) and an edge from each coloring v_n of $\{1, 2, \ldots, n\}$ to each coloring v_{n+1} of $\{1, 2, \ldots, n, n+1\}$ that agrees with v_n on the i-element subsets of $\{1, 2, \ldots, n\}$.

By assumption, the tree T is infinite. Hence T contains an infinite branch B, by the König infinity lemma. Also, since each coloring of $\{1, 2, \ldots, n\}$ is connected to a compatible coloring of $\{1, 2, \ldots, n, n+1\}$, B defines a coloring of the set of all the positive integers. So, by the infinite Ramsey theorem, the coloring defined by B has an infinite monochromatic subset.

But then, by following branch B to a sufficiently great depth n, we get a coloring of $\{1, 2, \ldots, n\}$ with a monochromatic subset of size m. This contradicts the choice of vertices in T, hence the finite Ramsey theorem is true.

direction from finite to infinite (as it was in the proof of compactness for propositional logic we gave in Section 4.7).

It happens that Ramsey proved his theorem to solve a problem in logic, but for several decades Ramsey theory was developed by combinatorialists, not logicians. Combinatorialists became interested in *Ramsey numbers*—the minimal values of N in the finite Ramsey theorem and certain special cases and variants of it. Given that N exists, it is computable from i, k and m by enumerating all k-colorings of the i-element subsets of sets with size m, $m+1$, $m+2, \ldots$ until we reach a set of size N whose k-colorings all have the required property. But the size of the computation grows rapidly with m and, because of this, few Ramsey numbers are actually known.[1]

Nevertheless, the value of N does not grow too rapidly for its existence to be unprovable in PA. Indeed, the finite Ramsey theorem (appropriately expressed as a sentence about positive integers) is a theorem of PA.

---

[1] For example, we do not know the minimal value of N such that a group of N people always contains five mutual acquaintances or five mutual strangers. It is known only that $43 \leq N \leq 49$.

Logicians became interested in Ramsey theory again in 1977, when Jeff Paris and the American logician Leo Harrington found a simply-stated variant of the finite Ramsey theorem that is *not* provable in PA. They call a finite set S of natural numbers *large* if the number of members of S is as great as the least member of S. Then their theorem is the following:

PARIS-HARRINGTON THEOREM. *For any i, k, and m, there is an N such that, for any $n \geq N$ and any k-coloring of the i-element subsets of $\{1, 2, 3, \ldots, n\}$, there is a large subset of size m whose i-element subsets all have the same color.*

With this variation in the statement of the finite Ramsey theorem, it turns out that N grows too rapidly (namely, at roughly the growth-rate of H_{ε_0}) for its existence for all i, k, m to be provable in PA. Thus the Paris-Harrington theorem in not provable in PA, though it is *expressible* in PA. It is also *true* (we would not be interested in it otherwise!) because it follows from the infinite Ramsey theorem. For more details, see Graham et al. (1990).

6.7   THE HARDEST THEOREM IN GRAPH THEORY

A *graph* (in the present context) G is essentially a finite set, whose members are called the *vertices* of G, and certain pairs of vertices, called the *edges* of G. That's all, though we should admit that one usually visualizes the vertices as dots, and the edges as lines connecting certain dots, as we have done already in discussing the graphs K_n in the last section. As we saw there, some simple-minded propositions about graphs can be quite hard to prove.

Another road to hard theorems in graph theory begins with the following easy theorem about positive integers: for any infinite sequence $n_1, n_2, n_3, \ldots$ of positive integers there are indices $i < j$ such that $n_i \leq n_j$. We can turn this result into a theorem of graph theory by replacing the positive integers $1, 2, 3, \ldots$ by the "line graphs" with $1, 2, 3, \ldots$ edges shown in Figure 6.4.

Figure 6.4. Line graphs corresponding to the positive integers.

With this correspondence between integers and graphs, the $\leq$ relation between integers corresponds to the *embedding* relation between line graphs. We say that graph G *embeds in* graph H if the vertices and edges of G may be mapped one-to-one and continuously onto the vertices and edges (respectively) of H. In this terminology, the easy theorem about integers becomes an easy theorem of graph theory: *for any infinite sequence $L_1, L_2, L_3, \ldots$ of line graphs there are indices $i < j$ such that L_i embeds in L_j.*

Now let's raise the degree of difficulty. Enlarge the class of line graphs to the class of *trees*, the finite graphs that are connected and contain no closed paths. The embedding relation makes sense for trees, but it is no longer obvious when one tree embeds in another. For example, in Figure 6.5, the first tree embeds in the second and third, but none of the other three trees embeds in any other.

Figure 6.5. Four trees.

Nevertheless, there is a theorem about infinite sequences of trees analogous to the above theorem about line graphs. It was discovered by the American mathematician Joseph Kruskal in 1960.

Kruskal's theorem. *For any infinite sequence of trees $T_1, T_2, T_3, \ldots$ there are indices $i < j$ such that T_i embeds in T_j.*

Another, perhaps more appealing, way to express Kruskal's theorem is the following: *any infinite sequence of distinct trees contains an infinite increasing subsequence in the embedding order.* Thus Kruskal's theorem is yet another case where "complete disorder is impossible."

The complexity of the embedding order for trees makes Kruskal's theorem quite hard to prove, and we will not even outline a proof here. (A proof may be found in the last chapter of the book by Diestel (2005).) What is remarkable is that being *apparently hard* to prove in this case corresponds to *actual unprovability* in systems such as PA. Of course, Kruskal's theorem cannot even be expressed in PA, because it involves arbitrary infinite sequences, so assumptions about infinity are surely needed for its proof. However, this is not really the source of the difficulty.

In 1981, the American logician Harvey Friedman found a natural *finite form* of Kruskal's theorem, expressible in PA, but still not provable in PA. In fact, the proof of Friedman's finite form requires induction up to an ordinal much larger than ε_0. The theorem is about finite sequences of trees, with a simple "rate of growth" condition that avoids obvious counterexamples:

Friedman's finite form of Kruskal's theorem. *For each m there is an n such that, if $T_1, T_2, \ldots, T_n$ is a sequence of trees in which T_k has at least mk vertices, then there are indices $i < j$ such that T_i embeds in T_j.*

This form of Kruskal's theorem implicitly involves a computable function, namely

$$f(m) = \text{least } n \text{ such that any such sequence } T_1, T_2, \ldots, T_n$$
$$\text{includes a } T_i \text{ that embeds in } T_j, \text{ for some } i < j.$$

The real source of difficulty in Kruskal's theorem is that f grows amazingly fast—not just faster than any provably computable function, but much faster than H_{ε_0}, the yardstick for our previous examples of unprovability in PA. Thus Friedman's discovery opened new vistas of unprovability—not only at a higher ordinal level, but arguably at a *lower* conceptual level. Explaining the concepts in Kruskal's theorem to a creature from outer space would probably be easier than explaining the concepts in Goodstein's theorem, for example.

In principle, finite graph theory can be expressed in PA, by coding vertices as numbers and edges as number pairs. But the *language* of graph theory seems better able to express hard theorems than the language of PA. Friedman's finite form of Kruskal's theorem is one example, and there are similar finite forms of the (even harder) *graph minor theorem* proved by the American mathematicians Neil Robertson and Paul Seymour between 1983 and 2004.

THE GRAPH MINOR THEOREM

The graph minor theorem is a variant of Kruskal's theorem in which trees are replaced by arbitrary finite graphs, and embedding by a looser concept called the *graph minor relation.* We have to relax the concept of embedding to get a Kruskal-type theorem for the infinite sequence of *polygon graphs* shown in Figure 6.6, no one of which embeds in a later member of the sequence.

The graph minor relation relaxes the concept of embedding just enough by allowing any vertex to *blow up* into a pair of vertices connected

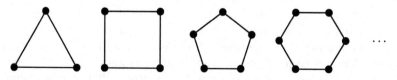

Figure 6.6. Polygon graphs.

by an edge. We call graph G' a *blowup* of graph G if G' results from G by a finite sequence of vertex blowups. For example, it is clear that any polygon graph is a blowup of any polygon graph with fewer vertices. (It is also true that any tree is a blowup of a single vertex; however, blowups cannot create new closed paths.)

Finally, we say that graph G is a *minor of* graph H if some blowup G' of G embeds in H. Then we have the following analogue of Kruskal's theorem for arbitrary graphs:

GRAPH MINOR THEOREM. *For any infinite sequence of graphs* $G_1, G_2, G_3, \ldots$ *there are indices* $i < j$ *such that* G_i *is a minor of* G_j.

Robertson and Seymour discovered the graph minor theorem around 1987, and proved it in a sequence of 20 papers completed only in 2004. It is by far the longest proof in graph theory to date, and it has not yet been compressed to a form suitable for textbooks (though important parts of it are given in the last chapter of Diestel (2005)). As with Kruskal's theorem, the apparent difficulty of the theorem is confirmed by proof theory.

Friedman found finite forms of the graph minor theorem which, like those he found for Kruskal's theorem, are expressible but not provable in PA. The finite versions of the graph minor theorem also have a level of unprovability higher than ε_0, though their exact level is not yet known.

6.8   HISTORICAL BACKGROUND

RAMSEY THEORY

The first theorems that we now consider to be part of Ramsey theory actually appeared a decade or two before Ramsey's work. They are theorems about the natural numbers of a type now called *arithmetic Ramsey theory*. A nice book on this branch of Ramsey theory is the one by Landman and Robertson (2004). The first two theorems of arithmetic Ramsey theory are the following theorems about finite sets of integers:

- SCHUR'S THEOREM (1916). *For any integer* $r \geq 1$ *there is an integer* s *such that, for any* r-coloring *of* $\{1, 2, \ldots, s\}$, *there is a monochromatic triple* (x, y, z) *such that* $x + y = z$.

- VAN DER WAERDEN'S THEOREM (1927). *For any integers* $k, r \geq 1$ *there is an integer* s *such that, for any* r-coloring *of* $\{1, 2, \ldots, s\}$, *there is a monochromatic arithmetic progression of length* k.

Ramsey himself, in his 1930 paper, proved the infinite Ramsey theorem, then gave a separate proof of the finite version. He did not point out the connection between them that we gave in Section 6.6. The realization that some kind of "compactness" relates finite and infinite Ramsey theorems occurs in König (1936), Chapter VI, where he shows that the König infinity lemma allows the finite van der Waerden theorem to be easily derived from an infinite one. The infinite van der Waerden theorem says that, in any r-coloring of the positive integers, at least one color contains arbitrarily long arithmetic progressions.

A remarkable strengthening of the infinite van der Waerden theorem was proved by the Hungarian mathematician Endre Szemerédi. Szemerédi (1975) proved that any set of positive integers of *positive density* contains arbitrarily long arithmetic progressions. The *density* of a set S of positive integers is the lower limit of the ratio

(number of members of S less than n)$/n$.

If the positive integers are colored in finitely many colors, it is clear that at least one color covers a set of positive density, so van der Waerden's theorem follows from Szemerédi's. However, Szemerédi's theorem is very difficult, and it has been the source of several new ideas in combinatorics.

An even more difficult theorem, building on Szemerédi's, was proved by the Australian Terry Tao and his English colleague Ben Green in 2004. The Green-Tao theorem solves a long-standing open problem by showing that *there are arbitrarily long arithmetic progressions in the prime numbers.* That is, there are arbitrarily large sets of "equally spaced" primes. Some small examples are $\{3,5,7\}$ and $\{5,11,17,23,29\}$, and the largest known example has 24 members (initial member 468395662504823 and distance 45872132836530 between successive members).

Thus the Green-Tao theorem proves the existence of sets that are extremely hard to find in practice. Its proof requires more firepower than Szemerédi's theorem, because the primes do *not* have positive density. Nevertheless, the associated function,

$f(m) =$ least n such that there are m equally-spaced primes less than n,

does not grow so fast as to rule out a proof of the Tao-Green theorem in PA. Tao has shown that f has the "slow" growth rate,

$$f(m) \leq 2^{2^{2^{2^{2^{2^{2^{2^{100m}}}}}}}},$$

so f is provably computable in PA.

In fact, as far as we know, all theorems of number theory not devised by logicians have proofs in PA. The Szemerédi and Green-Tao theorems have proofs that are simplified (even motivated) by infinitary methods, but at a pinch the infinitary methods can be eliminated. [2] This is not the case for the Paris-Harrington theorem, which was devised and proved by logicians. (See their proof in Paris and Harrington (1977).) Nevertheless, the condition that accelerates growth rate enough to ensure unprovability in PA—that each finite set have size greater than its smallest element—is natural enough. The theorem has now been embraced by combinatorialists, as its inclusion in the book of Graham et al. (1990) shows. It may be that combinatorics is actually more friendly to logic than number theory, in the sense that it more naturally expresses principles stronger than PA.

GRAPH THEORY

Graph theory is a cuckoo in the combinatorial nest.
—Peter J. Cameron (1994), p. 159.

The huge Robertson-Seymour theory of graph minors grew from a small seed known as *Kuratowski's theorem*, discovered in 1930 by the Polish mathematician Kazimierz Kuratowski. The theorem concerns *planar graphs*—those that can be drawn in the plane without edges crossing. An example of a planar graph is the *cube graph*, two views of which are shown in Figure 6.7. Since no edges cross in the first view, the cube graph is a planar graph. We also call the first view an *embedding* of the cube graph in the plane.

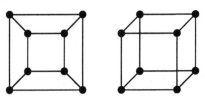

Figure 6.7. Two views of the cube graph.

Two important *non*planar graphs are shown in Figure 6.8. They are K_5, the complete graph on five vertices, and a graph called $K_{3,3}$, in which one set of three vertices is connected to another. $K_{3,3}$ is often known as the "utilities graph" after a popular puzzle in which three houses are supposed to be connected to three utilities (gas, water, electricity) by lines that do not cross. Because $K_{3,3}$ is nonplanar, the puzzle is unsolvable.

---

[2]In fact, it appears that all theorems of number theory proved by mainstream mathematicians can be proved in quite a small fragment of PA called EFA. In EFA, any provably computable function is bounded by a finite iterate of the exponential function. For a discussion of this surprising situation, see Avigad (2003).

Figure 6.8. The nonplanar graphs K_5 and $K_{3,3}$.

Experience, such as trying to solve the utilities puzzle, suggests that neither K_5 nor $K_{3,3}$ can be drawn in the plane without edges crossing. To prove this, we use a formula that goes back to 1752, the *Euler polyhedron formula*,

$$V - E + F = 2.$$

Here V is the number of vertices, E is the number of edges, and F is the number of faces (regions of the plane cut out by the edges) of a connected graph embedded in the plane.

NONPLANARITY AND THE EULER POLYHEDRON FORMULA

In the case of K_5 we have $V = 5$, $E = 10$, so the number of faces in any embedding of K_5 in the plane is necessarily $F = 7$. Each face has at least three edges (because in K_5 only one edge connects any two vertices), which makes a total of at least 21. Of course, this figure counts each edge twice, since each edge is in the boundary of two faces. But we still get more than 10 edges, which is a contradiction. So K_5 is nonplanar.

There is a similar proof that $K_{3,3}$ is nonplanar, using the fact that $K_{3,3}$ contains no triangles, hence any face of its hypothetical planar embedding has at least four edges.

This brings us, finally, to the theorem of Kuratowski (1930). It says that, in a certain sense, *any nonplanar graph "contains" a K_5 or a $K_{3,3}$.* There are at least two ways to interpret the word "contains":

KURATOWSKI'S VERSION (1930). *Any nonplanar graph has a subdivision of K_5 or $K_{3,3}$ as a subgraph.* (A *subdivision* of a graph G is the result of inserting new vertices in the interior of some edges of G.)

WAGNER'S VERSION (1937). *Any nonplanar graph has K_5 or $K_{3,3}$ as a minor, that is, it has a blowup of K_5 or $K_{3,3}$ as a subgraph.*

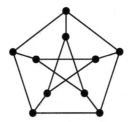

Figure 6.9. The Petersen graph.

Both versions identity K_5 and $K_{3,3}$ as the *forbidden subgraphs* for planar graphs, but in different ways. The difference may be seen from the graph in Figure 6.9, a nonplanar graph known as the *Petersen graph*.

The Petersen graph is obviously a blowup of K_5, but it does not have a subdivision of K_5 as a subgraph. Instead, it has a subgraph that is a subdivision of $K_{3,3}$. (It is not easy to find. Try looking for it!)

Extending the Kuratowski and Wagner theorems to other surfaces proves to be surprisingly difficult. A natural way to start is to classify surfaces by their *genus g* (number of "holes") and to find which K_n can be embedded in the surface of genus g. Figure 6.10 shows the first few of these, starting with the sphere (zero holes). The planar graphs are precisely those that may be embedded in the sphere.

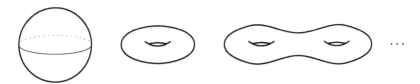

Figure 6.10. Surfaces of genus $0, 1, 2, \ldots$.

It is known that K_5, K_6, K_7 may be embedded in the surface with one hole (the torus) but not K_8. However, the theorem that answers the question for all K_n and all g was proved only in 1968, and the proof fills a whole book, Ringel (1974). The answer is that K_n embeds in the surface with g holes only for

$$n \leq \frac{1}{2} \left(7 + \sqrt{48g + 1} \right).$$

This theorem gives us *one* forbidden subgraph for graphs on the surface with g holes, namely K_m where m is the least integer greater than $\frac{1}{2} \left(7 + \sqrt{48g + 1} \right)$. As for other forbidden subgraphs, it is not even clear that the number is finite!

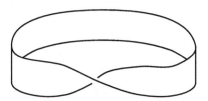

Figure 6.11. A Möbius band.

The first theorem, other than Kuratowski's, establishing a finite number of forbidden subgraphs was obtained for a surface not even mentioned yet—the *Möbius band*. The Möbius band is not in the above list of "surfaces with holes" because that list includes only the so-called *orientable* ("two-sided") surfaces, and the Möbius band is one-sided. You can make a Möbius band by joining the ends of a rectangle, with a twist—see Figure 6.11.

We can check that K_5 and K_6 embed in the Möbius band, but it turns out that K_7 does not. Thus K_7 is a forbidden subgraph for graphs on the Möbius band. Glover et al. (1979) proved that there are 103 (!) forbidden subgraphs for the Möbius band.[3] The number 103 results from taking the Kuratowski interpretation of forbidden subgraph; under the Wagner interpretation there are 35 forbidden subgraphs.

We do not yet know the number of forbidden subgraphs for graphs on the torus, but we now know that the number is finite! This, and the corresponding result for any surface, is a consequence of Robertson and Seymour's graph minor theorem. More precisely, it follows easily from the graph minor theorem that, for each surface S, there is a finite set of *forbidden minors*. A forbidden minor for S is a graph that does not occur as a minor of a graph embeddable in S, and which is minimal in the minor ordering. For example, K_5 does not occur as a minor of any planar graph, and it is minimal. K_6 also does not occur as a minor for any any planar graph, but it is *not* minimal (and hence not a forbidden minor) because it has K_5 as a minor.

It follows from this definition that, in the set of forbidden minors for any surface, no member is a minor of another. Therefore, by the graph minor theorem, *the set of forbidden minors is finite*. Nevertheless, the number of forbidden minors is still not known for any surface of genus greater than zero, and the number appears to grow very rapidly. It is known that there are at least 16000 forbidden minors for the torus.

---

[3]Strictly speaking, Glover, Huneke, and Wang proved their theorem for the *projective plane*, an abstract surface obtained from the Möbius band by spanning its boundary curve by a disk. This makes no difference as far as graph-embedding is concerned, but it is harder to draw because the projective plane cannot be embedded in ordinary 3-dimensional space.

COMBINATORICS AND ANALYSIS

These principles allow one to tap the power of the infinitary world (for instance, the ability to take limits and perform completions or closures of objects) in order to establish results in the finitary world, or at least to take the intuition gained in the infinitary world and transfer it to a finitary setting.
—Terry Tao (2009), p. 165.

In recent years, combinatorics has been increasingly influenced by ideas from analysis: real numbers, limits, real functions, infinite series. Examples are a new proof of Szemerédi's theorem in 2002 by the English mathematician Tim Gowers, and the Green-Tao proof that there are arbitrarily long arithmetic progressions in the primes in 2004. As Terry Tao claims, *analysis gives intuition about the infinitary world*, which may be used to establish results in the finitary world. In this book, we have now seen many examples of results about the finitary world that depend on infinitary assumptions. However, not many of these infinitary assumptions are typical of analysis. Here is a simple example of one that is, involving a classical theorem of analysis.

The *Bolzano-Weierstrass theorem* states that any infinite set S in the unit interval $[0, 1]$ has a limit point. That is, there is a point X, each neighborhood of which includes points of S. We mentioned this theorem once before, in connection with the related Heine-Borel theorem and the König infinity lemma, and indeed the two results have similar proofs. However, the Bolzano-Weierstrass theorem is arguably more intuitive, because we are more familiar with sets of points in the unit interval than with infinite trees.

Now consider the following theorem, reminiscent of the Kruskal theorem about trees, but more elementary: *every infinite S sequence of distinct numbers in $[0, 1]$ contains either an infinite increasing subsequence, or an infinite decreasing subsequence.* This theorem can certainly be proved by a combinatorial argument—for example, by coloring increasing pairs red and the other pairs blue it follows from the infinite Ramsey theorem— but it is *immediate* from the Bolzano-Weierstrass theorem!

Consider a limit point X of the set of members of S. Either there are infinitely many members of S less than X, in which case we can find an infinite increasing subsequence (with limit X), or there are infinitely many members of S greater than X, in which case we can find an infinite decreasing subsequence (with limit X).

While this theorem is only a baby theorem of Ramsey theory, I think that it illustrates the power of intuition from analysis. Analysis allows us to create objects, such as limit points, that realize infinite sets or processes, and allow us to visualize them. The results of Gowers, Green, and Tao show this power more convincingly, but unfortunately they are far beyond the scope of this book.

A related branch of mathematics that has been applied to combinatorics is *ergodic theory*. Roughly speaking, ergodic theory studies "behavior in the limit" of infinite processes taking place in spaces such as
the line or plane. In the 1970s, the Israeli mathematician Hillel Furstenberg gave simplified proofs of Szemerédi's theorem (1977) and van der
Waerden's theorem (1978) by relocating them in ergodic theory. Amazingly, there is a structural connection between Furstenberg's proof and
the original proof by van der Waerden. The French logician Jean-Yves
Girard showed in his book Girard (1987) that when Furstenberg's proof
is suitably expanded, by cut elimination, it becomes van der Waerden's
proof!

Girard's discovery shows by example how infinitary principles of
analysis can "repackage" infinitary principles of combinatorics so as to
make them familiar to mainstream mathematicians. For a survey of what
logic currently has to say about ergodic theory, see Avigad (2009).

- CHAPTER 7 -

AXIOMS OF INFINITY

PREVIEW

The adaptation of strong axioms of infinity is thus a theological venture, involving basic questions of belief concerning what is true about the universe. However, one can also construe work in the theory of large cardinals as formal mathematics, that is . . . those formal implications provable in first order logic.

—Kanamori and Magidor (1978), p. 104.

This chapter is in the nature of an epilogue, tying up several threads from earlier chapters of the book: the nature of infinity, how it limits the scope of provability, and how it affects our knowledge of the natural numbers.

The roads to infinity discovered by Cantor—via the diagonal construction and via the countable ordinals—turn out to be just two of many ways to comprehend the vast universe of uncountable sets. By the 1930s, mathematicians were contemplating sets large enough to model the universe itself: the so-called *inaccessible* and *measurable* cardinals.

We observe the mysterious influence of these incomprehensibly large sets on the properties of real numbers, before returning to arithmetic, where the influence of these incomprehensible infinities can also be detected. In fact, it becomes clear that a better understanding of finite objects depends on a better understanding of infinity.

7.1 SET THEORY WITHOUT INFINITY

When people speculate about alien intelligence, and how we might learn of its existence, they usually think about numbers. The sequence of integers

$$1, 2, 3, 4, 5, 6, 7, 8, 9, \ldots,$$

or the Fibonacci numbers

$$1, 2, 3, 5, 8, 13, 21, 34, 55, \ldots,$$

or the primes

$$2, 3, 5, 7, 11, 13, 17, 19, 23, 29, 31, \ldots,$$

are "messages" that we would immediately recognize as signs of intelligence. But what if the language of alien math was entirely different from ours?

We have already had a taste of "alien number theory" when we noted von Neumann's definition of the natural numbers in Section 1.8. Taking the empty set to be 0, von Neumann defines the successor function by

$$S(n) = n + 1 = n \cup \{n\}.$$

This definition also gives the "less than" relation in the simplest possible way: $m < n$ if and only if m is a member of n, written $m \in n$. As we know, sum and product are definable inductively from the successor function, because

$$m + 0 = m, \quad m + S(n) = S(m + n),$$
$$m \cdot 0 = 0, \quad m \cdot S(n) = m \cdot n + m.$$

Thus, all we really need to extract Peano arithmetic (PA) from set theory is induction. This is easy too, using the ZF axioms listed in Section 1.8. We want to prove that, for any definable property $\mathcal{P}$, if there is a natural number with property $\mathcal{P}$ then there is a *least* natural number with property $\mathcal{P}$. Before we explain why this is possible, a couple of points need to be clarified.

- We need to define the property "n is a natural number" in the language of ZF set theory. Thanks to von Neumann's definition of natural number, this property is a combination of properties involving membership:

  - for any distinct $p, q \in n$, either $p \in q$ or $q \in p$.
  - any member of a member of n is a member of n.
  - if n is not 0, then n has a greatest member.

- We need to show that any property $\mathcal{P}$ definable in PA is definable in ZF set theory. This follows because the basic concepts of PA are 0 and successor, both of which are definable in ZF, as we have seen.

Now, if n is a natural number with property $\mathcal{P}$, consider the numbers less than n, which are members of n. The members with property $\mathcal{P}$ form a set (thanks to the replacement axiom, Axiom 6 on the list in Section 1.8). So it remains to show that any nonempty set has a least member with respect to the membership relation. This is an axiom! (The axiom of foundation, Axiom 7 on the list in Section 1.8.)

Thus PA is expressible in ZF set theory. Notice also that *we have not used the axiom of infinity* (Axiom 8 on the list in Section 1.8) to build the concepts and axioms of PA. So PA is in fact expressible in a weaker set theory: ZF minus the axiom of infinity, or ZF − Infinity for short.

We can say that ZF − Infinity is weaker than ZF, because the axiom of infinity is definitely not provable from the remaining seven axioms (unless they are inconsistent). These seven axioms have a model consisting of the so-called *hereditarily finite* sets, the sets built from the empty set by the pairing, union, power, and replacement operations. Any one of these operations on a finite set obviously produces a finite set, hence all sets in the model are finite. (Moreover, all members of such a set are finite, all members of members, and so on, which is why these sets are called *hereditarily* finite.) The hereditarily finite sets satisfy the first seven axioms of ZF, by construction, but they do not satisfy the axiom of infinity. Hence the axiom of infinity does not follow from the others.

In fact, since ZF − Infinity can be modeled by the hereditarily finite sets, *it can be encoded in PA*. In PA, as we saw in Section 3.3, we can define a concatenation relation that encodes a sequence of positive integers by a single integer. By more intensive use of concatenation, we can in fact encode all hereditarily finite sets by positive integers; also the various finite operations on sets. This encoding shows that *ZF − Infinity and PA are essentially the same*. ZF − Infinity is an alien number theory!

No doubt most humans would prefer to do number theory in PA rather than in ZF − Infinity, but it is good to be aware of the alternative. At the very least, ZF − Infinity is a good setting for areas of mathematics based directly on finite sets, such as Ramsey theory and graph theory. As the barriers between number theory and these areas melt away, ZF − Infinity may not seem alien after all. It may be seen rather as the natural base theory of mathematics, from which "higher mathematics" is obtained by adding various axioms of infinity.

What a Difference Infinity Makes

Now that we know that ZF − Infinity is essentially PA, we can see what an enormous difference the axiom of infinity makes. From a universe in which only the hereditarily finite sets need exist, the axiom of infinity takes us to a universe in which a set of size $\aleph_0$ exists. Then the power set axiom gives us sets of size $2^{\aleph_0}$, $2^{2^{\aleph_0}}$, ... and the replacement axiom gives others (which may be different) such as $\aleph_1, \aleph_2, \ldots$. Since we have a function $n \mapsto \aleph_n$ whose domain is the set ω, the axiom of replacement gives us the set $\{\aleph_0, \aleph_1, \aleph_2, \ldots\}$. Applying the union axiom to this set gives $\aleph_\omega$, the least upper bound of $\aleph_0, \aleph_1, \aleph_2, \ldots$, and so it goes, onward and upward to $\aleph_\alpha$ for any ordinal α.

Even at the level of PA, adding the axiom of infinity makes a big difference to what can be proved. The set $\{0, 1, 2, \ldots\}$, whose existence follows from the axiom of infinity, satisfies the PA axioms, and hence proves that PA is consistent. Thus Con(PA), which we know from Gödel incompleteness is *not* a theorem of ZF − Infinity (unless PA is inconsistent), becomes a theorem when the axiom of infinity is added. It is the same for ε_0-induction, which is provable in ZF by developing the theory of ordinals.

7.2 Inaccessible Cardinals

Of course, the sky looks highly unapproachable from the ground.
—Kanamori and Magidor (1978), p. 115.

The universe V arising from the axiom of infinity is "measured" by ordinals, in the sense that each set belongs to an initial segment V_α of V consisting of sets built by $< \alpha$ applications of the power set operation. Thus V_0 is the empty set; $V_{\alpha+1}$ is the power set of V_α; and V_λ, for limit λ, is the union of the V_β for $\beta < \lambda$. The least α for which a set X appears as a subset of V_α is called the *rank* of X. The axiom of foundation implies that every set has a rank.

Consider the set V_ω of sets of rank $< \omega$. The members of V_ω are the sets of all finite ranks: the hereditarily finite sets, which we saw in the previous section form a model of ZF − Infinity. If we go to some higher rank α, which necessarily includes infinite sets, is it possible to model *all* the axioms of ZF by V_α? If so, V_α must be closed under the power set and replacement operations. This is a tall order for a set with infinite members. None of the ordinals α we can name—such as $\aleph_0, \aleph_1, \aleph_2, \ldots, \aleph_\omega$—give a V_α with these closure properties.

A set V_α that is closed under the power set and replacement operations is necessarily of the form V_κ for some cardinal number κ. The cardinal κ is then called *inaccessible*, because in some sense V_κ cannot be "reached from below." The cardinal ω is inaccessible in this sense, but *in ZF we cannot prove the existence of an inaccessible $\kappa > \omega$*. This is because we could use V_κ to model all the axioms of ZF, and hence prove in ZF the sentence Con(ZF) asserting that ZF is consistent. But ZF includes PA, hence *we cannot prove* Con(ZF) *in* ZF, by Gödel's second incompleteness theorem.

Thus, the assertion that an inaccessible $\kappa > \omega$ exists is a *new kind of axiom of infinity*, stronger than the ordinary axiom of infinity. Incidentally, this shows that unprovable sentences of ZF are relatively natural and easy to think up. (Indeed, inaccessibles *were* thought up, long before any sentence of ZF was known to be unprovable.) Any sentence implying the existence of inaccessibles, such as existence of cardinals *larger* than inaccessibles, is necessarily unprovable in ZF (unless ZF is inconsistent).

INACCESSIBLES AND MEASURABILITY

> Icarus-like, soar unashamedly in speculative altitudes implicit in forms of statements like "if there is a measurable cardinal, then ...". If the sun begins to melt the wax on our wings, we can always don our formalist parachute by saying that these are interesting implications .... .
> —Kanamori and Magidor (1978), p. 115.

Since so many sentences of ZF are unprovable, one becomes interested in proving *relative consistency statements*, of the form:

If ZF + Axiom A is consistent, then ZF + Axiom B is consistent.

Examples we have already mentioned are those proved by Gödel and Cohen concerning the axiom of choice (AC) and the continuum hypothesis (CH):

- If ZF is consistent then ZF + AC + CH is consistent (Gödel (1938)).

- If ZF is consistent then ZF + ¬AC is consistent (Cohen (1963)).

- If ZF is consistent then ZF + AC + ¬CH is consistent (Cohen (1963)).

One of the most remarkable relative consistency statements was proved in 1964 by Robert Solovay, one of the post-Cohen wave of American set theorists. Solovay proved that:

If ZF + AC + "an inaccessible exists" is consistent,

then ZF + "every set of reals is Lebesgue measurable" is consistent.

Twenty years later, the Israeli mathematician Saharon Shelah showed that the existence of an inaccessible is really essential to Solovay's result— assuming only that ZF + AC is consistent is not strong enough to prove the consistency with ZF of having all sets of reals Lebesgue measurable.

If all sets of reals are Lebesgue measurable then, in particular, all subsets of the unit interval $[0, 1]$ have measure between 0 and 1. It follows, when we identify the cardinal number $2^{\aleph_0}$ with the set of points in $[0,1]$ that *the cardinal number* $2^{\aleph_0}$ *is measurable*, in the sense that every subset of $2^{\aleph_0}$ can be assigned a real number, called its *measure*, with the following properties:

- $2^{\aleph_0}$ has measure 1.

- Each 1-element subset has measure 0.

- The union of countably many disjoint subsets, with respective measures $m_1, m_2, m_3, \ldots$, has measure $m_1 + m_2 + m_3 + \cdots$.

This raises the questions: what is an appropriate definition of "measurable" for other cardinal numbers, and do any "measurable" cardinals exist?

These questions have a long history, beginning with the Polish school of Stefan Banach in the late 1920s, and we refer to Kanamori (2003) for details. In short, there is an appropriate definition of measurable cardinals, but their existence cannot be proved because it implies the existence of inaccessibles. Like the existence of inaccessibles, the existence of measurable cardinals is an axiom of infinity, and it implies many interesting results not provable in ZF alone. The sets asserted to exist in such axioms— namely, sets larger than any whose existence can be proved from the ZF axioms—are often known simply as *large cardinals*.

Kanamori (2003) is the most complete source of information about large cardinals. Long stretches of it are unavoidably technical, but it contains some highly readable and well-informed passages on the history of set theory.

7.3 The Axiom of Determinacy

> If axioms postulating the existence of very large cardinals are difficult to motivate ... then why have they been studied so thoroughly? The simple reason is that they lead to substantial and beautiful mathematics.
>
> —Robert Wolf (2005), p. 254.

In 1925, Stefan Banach and his colleague Stanisław Mazur started a new and curious thread in the study of real numbers by considering a game in which players I and II take turns to pick alternate decimal digits of a number x. The game proceeds for ω steps, the aim of player I being to keep x in some set X given in advance, and the aim of player II being to keep x out. For example, if X is the set of rational numbers, then II can win by playing some nonperiodic sequence of digits, such as

12112111211112.... . No matter which digits are played by I, the number x has a nonperiodic decimal expansion, and hence is irrational. Thus *player II has a winning strategy for the Banach-Mazur game when X is the set of rational numbers.*

A set X of real numbers is said to be *determined* when one of the players has a winning strategy in the Banach-Mazur game for X. Using the axiom of choice, Banach and Mazur were able to show that there is a set X which is not determined. This put the question of determinacy back on the shelf for nearly 40 years. Although no non-determined examples were found without the axiom of choice, there was also very little progress in finding determined examples. By the early 1960s the only positive results were that sets up to the third level of the Borel hierarchy are determined.

Then, in 1964, the two Polish mathematicians Jan Mycielski and Stanisław Swierczkowski made a find that put determined sets in an entirely new light: *all determined sets are Lebesgue measurable.* Suddenly, determinacy was a desirable property, so much so that one might want an *axiom of determinacy,* stating that all sets of real numbers are determined. The axiom of determinacy implies that the axiom of choice is false, so it cannot be proved in ZF. But it is an attractive *alternative* to the axiom of choice, because sets of reals are better behaved (at least as far as measurability is concerned) under the axiom of determinacy than under the axiom of choice.

The proposal of the axiom of determinacy (AD) initiated a 20-year search to find a basis for it in some axiom of infinity, a search involving some of the deepest and most intricate work in the history of set theory. The main results were obtained by the American mathematicians Donald A. Martin, John Steel, and Hugh Woodin. Their odyssey began with the result of Martin (1969/1970) that, assuming the existence of a measurable cardinal, *analytic* sets are determined. The analytic sets are the projections (on the line) of all Borel subsets of the plane. They include all Borel subsets of the line and also some sets that are not Borel, so Martin vastly extended the known world of determined sets.

Unfortunately, not all sets are analytic—for example, the complements of some analytic sets are not analytic—so Martin's result falls short of proving determinacy for any reasonably extensive class of definable sets. And, as we have said, Martin had to assume the existence of a measurable cardinal. (In 1974 he was able to prove determinacy of Borel sets in ZF + AC alone, but it is now known that determinacy of analytic sets needs a large cardinal, because with ZF + AC it actually *implies* the existence of large cardinals.)

One does not expect to prove the full AD, even from extravagantly large cardinal assumptions, because it contradicts the axiom of choice.

However, it would be nice to know that AD is *consistent* with ZF. To prove this we need an axiom of infinity much stronger than just the existence of measurable cardinals. The corresponding large cardinals were introduced by Woodin in 1984 and are called *Woodin cardinals*. If they exist, they are much larger than the smallest measurable cardinal. And to prove the consistency of AD, we need to assume that there are *infinitely many* Woodin cardinals. With this massive assumption, together with the axiom of choice, Woodin proved in 1985 that:

ZF + AD is consistent if and only if

ZF + AC + "there are infinitely many Woodin cardinals" is consistent.

Thus AD is an axiom of infinity, in the same sense (but more so) as the assumption that all sets of reals are Lebesgue measurable.

The greater strength of Woodin cardinals can also be seen in the amount of determinacy they directly imply. Measurable cardinals imply the determinacy of analytic sets, but not much more. It is known, for example, that they do not imply determinacy for projections of complements of analytic sets. Woodin cardinals, on the other hand, imply determinacy for *all* sets obtained from analytic sets by a finite number of complementations and projections—the so-called *projective sets*. This was shown by Martin and Steel in 1985. To many set theorists, this seems like a nice place to erect a fence shutting out the indeterminate sets. It allows all projective sets to have nice properties, such as measurability, and the axiom of choice can still hold. But the nonmeasurable sets entailed by the axiom of choice are outside the class of the projective sets, and hence are remote from classical mathematics.

7.4 Largeness Axioms for Arithmetic

We have seen in Chapters 3 and 5 how unprovability in PA was discovered by "miniaturization" of set theory arguments. Gödel's "miniature" diagonal argument leads to the unprovability of Con(PA), and Gentzen's "miniaturization" of the proof that there are uncountably many countable ordinals leads to the unprovability of ε_0-induction. We might call ε_0-induction a "large ordinal axiom" for PA. Like the large cardinal axioms for ZF, it implies some interesting sentences that are not otherwise provable, such as Goodstein's theorem.

Another way to express "largeness" in PA is via propositions that imply the existence of rapidly growing functions. Goodstein's theorem is of this type, as is the Paris-Harrington theorem. Friedman's finite form of Kruskal's theorem expresses greater "largeness" than the Goodstein and

Paris-Harrington theorems, because it implies α-induction for a countable ordinal α much larger than ε_0. Thus there is potentially a plethora of largeness axioms for PA, just as there is already for ZF.

However, the landscape of such axioms for PA is much less explored than the landscape of large cardinal axioms for ZF, and we know few interesting sentences of PA implied by largeness axioms. In comparison, ZF has dozens of interesting unprovable propositions, many of which are settled by large cardinal axioms. The number of large cardinal axioms is itself "large." The book Kanamori (2003), p. 472, contains a full-page chart of no less than 29, with the known implications between them.[1] It remains to be seen whether largeness axioms expressible in PA will ever be as fruitful as axioms of infinity in ZF, though chances are good if PA continues to develop in the direction of Ramsey theory and graph theory. This may depend on mathematicians getting used to interpreting PA as ZF − Infinity.

Looking further afield, axioms of infinity *not* expressible in PA, such as the existence of inaccessibles, can certainly have consequences in PA that are not otherwise provable. For example, Con(ZF) is expressible in PA, but it is not provable in PA because it implies Con(PA). However, Con(ZF) is provable in ZF + Inaccessible, as we have seen in Section 7.2. Whether there are consequences of inaccessibles in PA as natural as the consequences of ε_0-induction is not yet clear.

7.5   LARGE CARDINALS AND FINITE MATHEMATICS

> It seems to us that the role of set theory in such cases is quite similar to the role of physics when the latter gives heuristic evidence for some statements that mathematicians are to prove subsequently... we think that the braid order is an application of set theory, and, more precisely, of this most speculative part of set theory that investigates large cardinals.
>
> —Patrick Dehornoy (2000), p. 600

Set theory is rarely compared with physics, but Patrick Dehornoy has a point, because physical intuition is a gift from the infinitary world. After all, our intuition of the continuum comes from the physical concepts of space and time, and we have known since Cantor that this is intuition about *uncountability*.

Dehornoy made a discovery in the world of finite objects, namely, the *linear ordering of braids*, based on his intuition about an object from

---

[1]The top "axiom" on Kanamori's chart is "$0 = 1$," reflecting the fact that some proposed large cardinal axioms have turned out to be false. This is a risk one runs when pursuing axioms that imply more and more sentences of ZF. A false axiom is maximally "strong" in the sense that it implies every sentence, but one does not want axioms to be this strong!

Figure 7.1. A typical braid.

the world of (very) large cardinals, namely, *existence of a self-similar V_α*. To understand how one bridges the gap between these two worlds one needs to read the article Dehornoy (1995) or the book Dehornoy (2000), but we can at least describe the two worlds at the ends of Dehornoy's remarkable bridge.

First, the world of braids. A *braid* is an arrangement in 3-dimensional space of disjoint curves, called *strings*. For example, the arrangement shown in Figure 7.1 is a 3-string braid, similar to the braiding used for hair.

Its characteristic features, which define braids in general, are these:

- The n threads connect the points $(0,0,1),(0,0,2),\ldots,(0,0,n)$ with the points $(k,0,1),(k,0,2),\ldots,(k,0,n)$ for some positive value k.

- Each thread is without "backtracking." That is, as one travels along a thread from initial point to final point the x-component of velocity is always positive.

- The projection of the braid on a vertical plane has only finitely many multiple points, and each multiple point is a *crossing* of one thread by another.

Because of these characteristics, each braid can be represented by a 2-dimensional diagram (as we have already done in Figure 7.1). The *braid diagram* is basically the projection of the braid on a vertical plane, except that each crossing is drawn so as to show which thread is nearest to the viewer. Two braids are considered to be the *same* if they differ by a deformation of the threads that leaves the initial and final points fixed and keeps the threads disjoint.

It is not obvious how one recognizes whether two given braids are the same, because deformation may alter the position and number of crossings. However, one can clarify this problem by deforming the braid diagram to a standard form, in which no two crossings have the same x coordinate. Then all the essential information about the braid is contained in the sequence of crossings in, say, left to right order. For example, the hair braid in Figure 7.1 is described by the sequence

top over middle,
bottom over middle,
top over middle,
bottom over middle,
top over middle.

Two braids $\mathcal{A}$, $\mathcal{B}$ with the same number of threads have a *product* $\mathcal{AB}$: the result of joining $\mathcal{A}$ and $\mathcal{B}$ together as shown in Figure 7.2. For each n, one has the *n-string braid group* B_n, consisting of all the n-string braids under the braid product operation. The problem of deciding when two n-string braids are the same may be solved by studying the algebraic structure of B_n, as was first done by Artin (1925). More recently, it has become common to investigate all the groups B_n simultaneously by merging them into a group B_∞. Each braid $\mathcal{A}_\infty$ in B_∞ has infinitely many strings, but only finitely many crossings, so $\mathcal{A}_\infty$ behaves like a braid in some B_n—namely, the braid $\mathcal{A}$ with enough strings to realize all the crossings of $\mathcal{A}_\infty$.

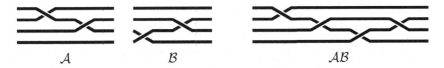

$$\mathcal{A} \qquad\qquad \mathcal{B} \qquad\qquad \mathcal{AB}$$

Figure 7.2. Two braids and their product.

Study of B_∞ was greatly assisted by Dehornoy's discovery, in 1992, that *there is a left-invariant linear order on the braids in B_∞*. That is, there is a linear order $<_B$ on the braids for which

$$\mathcal{A} <_B \mathcal{B} \quad \text{implies} \quad \mathcal{CA} <_B \mathcal{CB} \quad \text{for any } \mathcal{C} \text{ in } B_\infty.$$

Since the braids in B_∞ are essentially finite objects, Dehornoy's theorem is about finite objects. In particular, it can be encoded as a theorem of PA. Yet he discovered the theorem through a study of large cardinals, and he insists that this is the natural path to it, even though lower-level proofs are now known. In Dehornoy (1995), p. 33, he writes,

> More precisely we would like to explain how the set theoretical intuitions somehow *had to* lead these results, and in particular why the introduction of an 'exotic' distributive operation and of a linear ordering on braids is actually a natural development of an investigation initiated a decade ago in set theory.

We have already said something about the investigation that Dehornoy refers to. Namely, it was the attempt to understand projective determinacy through large cardinal axioms, which led to the results of Woodin, Martin, and Steel asserting the equivalence of projective determinacy with the existence of infinitely many Woodin cardinals.

In the course of searching for the minimal large cardinal assumption that implies projective determinacy, other large cardinal assumptions were studied. One of these, the existence of a self-similar V_α, ultimately proved stronger than needed for projective determinacy, but it was good for other purposes. In particular, it was the first assumption known to imply braid ordering.

Recall from Section 7.2 that V_α is the set of sets of rank α, that is, all sets that can be constructed by α applications of the power set operation. We say that V_α is *self-similar* if there is a map f of V_α into itself, not fixing all members of V_α, but preserving all properties that can be defined in the language of ZF. Such properties include, for example, "X is empty," "X is a member of Y," "X is an ordinal," "X is a successor," and so on. It follows that the map f *does* fix almost any set one can think of. For example

$f(0) = 0,$ because 0 is the unique empty set,

$f(1) = 1,$ because 1 is the unique successor of 0,

$$\vdots$$

$f(\omega) = \omega$ because ω is the unique set with members $0, 1, 2, \ldots$.

In fact, it can be shown that the least ordinal *not* fixed by f is greater than infinitely many Woodin cardinals, so the existence of a self-similar V_α is a large cardinal axiom, even stronger than the existence of infinitely many Woodin cardinals.

As we have said, this particular large cardinal axiom is interesting because it led to the *discovery* of the braid ordering. We now know that braid ordering can be proved from weaker assumptions, such as ZF. However, it is not always possible to eliminate large cardinal assumptions from theorems about finite objects. In a research program that is still under way, Harvey Friedman has found very large cardinal axioms with very finite consequences that are not otherwise provable. He has found many examples where the consequences are in PA. It is debatable whether his examples are as "natural" as the braid ordering, but they nevertheless make the point that *there are necessary uses of infinity, even in finite mathematics.*

7.6 Historical Background

Axioms of infinity

The existence of one such number [an inaccessible cardinal] ...seems at least
problematic; however, it must be considered as a possibility in what follows.
—Felix Hausdorff (1908), p. 444.

The observation that ZF − Infinity is essentially PA is due to Hilbert's
student Wilhelm Ackermann. In 1937 he proved the consistency of ZF −
Infinity by encoding it in PA, then appealing to Gentzen's consistency
proof for PA. He also showed that PA may be encoded in ZF − Infinity,
so the two theories are equiconsistent.

Inaccessible cardinals were first considered by Felix Hausdorff in 1908,
and studied more thoroughly by Paul Mahlo in 1911–1913. Mahlo considered limits of increasing sequences of inaccessibles, now known as *Mahlo
cardinals*. Hausdorff and Mahlo actually considered a concept now called
weak inaccessibility, which agrees with ordinary inaccessibility under the
so-called *generalized continuum hypothesis*: $2^{\aleph_\alpha} = \aleph_{\alpha+1}$. This hypothesis
was also introduced by Hausdorff in his 1908 paper, so in a way he con-
sidered ordinary inaccessibility too.

The stronger concept of inaccessibility, involving closure under both
the replacement and power operations, was introduced by Sierpiński and
Tarski in 1930. They also introduced the term "inaccessible" for the concept, attributing it to Kuratowski. In a remarkable paper, Kuratowski
(1924) pointed out (without proof, and assuming consistency) that the
existence of inaccessibles is not a theorem of ZF + AC. Namely, *if inaccessibles exist, then their existence is not provable in ZF + AC*—an exquisitely
theological predicament. However, Kuratowski was using the weak inaccessible concept of Hausdorff, so his claim could not have been proved at
that time without assuming the generalized continuum hypothesis. This
obstacle was pointed out by Kanamori (2003), p. 18. Kanamori credits
the first proof that the existence of inaccessibles is unprovable to Zermelo
(1930).

In fact, Zermelo presented his result in a talk at Freiburg on Febru-
ary 2, 1928, and it was reported in Baer (1928), p. 383. Zermelo's proof
would not be considered rigorous today, but his idea is so exquisitely
simple that the proof can be easily rewritten to comply with current standards. It does not involve anything as sophisticated as Gödel's incompleteness theorems, the second of which also shows that the existence of
inaccessibles cannot be proved in ZF + AC.[2]

---

[2]Indeed, Zermelo failed to understand Gödel's use of the diagonal argument and did
not believe that Gödel's theorems were correct. He still did not get the idea even after Gödel
wrote him a long letter in 1931, which may be found in Gödel (2003), p. 423. This perhaps
confirms that Zermelo's incompleteness theorem is simpler than Gödel's.

Zermelo's incompleteness theorem

If inaccessible κ exist, then this fact is not provable in ZF + AC. This is because $V_κ$ is a model of ZF+AC. So, if κ is the *smallest* ordinal with this property, then $V_κ$ satisfies

ZF + AC + "there is no inaccessible,"

because the α in $V_κ$ are such that $α < κ$, hence they are not inaccessible.

Also in 1930, Kuratowski's student Stanisław Ulam found that measurable cardinals are inaccessible. (Because of this, Kuratowski described Ulam as "my greatest discovery.") But it was not until 1960 that measurable cardinals were found to be larger than the smallest inaccessible cardinal—by Tarski and his student William Hanf—and hence of independent interest.[3] It follows that, if measurable cardinals exist, one cannot *prove* their existence even in ZF + AC + Inaccessible. This is because there is a smallest measurable cardinal κ, and then $V_κ$ satisfies not only ZF + AC + Inaccessible but also "there is no measurable cardinal," since all the cardinals in $V_κ$ are less than κ. (The cardinals in $V_κ$ are also nonmeasurable *from the viewpoint of* κ, though this not obvious.)

After this, the world of large cardinals expanded rapidly. A fine account of this world, and its history, is given in Kanamori (2003). Really intrepid readers may wish to consult Jech (2003) for further technical details, such as Solovay's theorem on the consistency of "all sets of reals are Lebesgue measurable."

Descriptive Set Theory

Lebesgue's trivial (as I have called it) mistake proved to be very useful: it unleashed an avalanche of new concepts and theorems.
—Kazimierz Kuratowski (1980), p. 70.

In the beginning, set theory was mainly a "foundational" discipline that aimed to clarify the basic concepts of analysis: numbers, functions,

---

[3]This time it was Gödel's turn not to understand. In a letter to Tarski that he drafted, but probably did not send, in August 1961 he wrote (see Gödel (2003), p. 273),

I have heard it has been proved that there is no two valued denumerably additive measure for the first inacc. number. I still can't believe that this is true ... .

limits, measure. Of course, Cantor quickly encountered a problem that the analysts had not thought of—the continuum hypothesis (CH). But there was reason to suppose that CH would be solved by the same idea that analysis seemed to need: a *classification* of sets of real numbers in terms of the *complexity* of their definitions. The theory developed to attack this classification problem came to be known as *descriptive set theory*.

We saw in Section 2.9 that by 1905 Borel, Baire, and Lebesgue had developed a classification of the Borel sets into $\aleph_1$ levels, partly as a means to describe the measurable sets. This was the first substantial result in descriptive set theory. Cantor himself did not get far in classifying sets of reals, but he introduced a property of sets that greatly influenced the future evolution of descriptive set theory—the *perfect set property*.

A set is called *perfect* if it is nonempty, closed, and contains no isolated points. For example, the interval $[0, 1]$ is a perfect set, but the set of integer points is not. It is quite easy to show that any perfect set has cardinality $2^{\aleph_0}$. Cantor (1883) proved that every uncountable closed set contains a perfect set, so any uncountable closed set has cardinality $2^{\aleph_0}$. We now express Cantor's result by saying that *the class of closed sets has the perfect set property*. More generally, we say that *a class S of sets of reals has the perfect set property* if any uncountable set X in S contains a perfect set. It follows that "S satisfies CH," in the sense that any uncountable member X of S has cardinality $2^{\aleph_0}$.

The closed sets lie at the very bottom of the Borel hierarchy, so Cantor's result is a long way from a proof of CH. However, one might hope to show that the perfect set property extends all the way up the Borel hierarchy, just as measurability extends all the way. In fact, all Borel sets *do* have the perfect set property, but this was established only in 1916, by Pavel Aleksandrov, a member of the Russian school headed by Nikolai Luzin.

Luzin studied in Paris in 1905 and was influenced by Borel and Lebesgue. At first his interests were in analysis, but by 1916 he had realized that his true calling was in descriptive set theory. His school gained fame through the discovery of his student, Mikhael Suslin, that the projections of Borel sets are *not* always Borel (the mistake in Lebesgue's 1905 paper). This brought to light a whole new class of sets: the projections of Borel sets, later known as the analytic sets. With them came the challenge of extending the results already known for Borel sets, such as measurability and the perfect set property.

In 1917 Luzin proved that analytic sets are measurable, and Suslin proved that they have the perfect set property, so Cantor's long march towards the continuum hypothesis was taken a step further. What was next? In 1925, Luzin and Wacław Sierpiński foreshadowed a move beyond the class of analytic sets by considering *all* sets obtainable from

Borel sets (of any dimension) by projection and complementation—what they called the projective sets.

The projective sets form a hierarchy of ω levels, originally called

A and CA   (analytic sets, and their complements),

PCA and CPCA   (projections of CA sets, and their complements),

and so on. (Today there is a shorter notation, but it takes longer to explain.) There are new sets in each level, but the task of proving measurability and the perfect set property seems insuperable even at the second level. Indeed, Luzin (1925) made the bold and prophetic declaration that "one does not know and *one will never know*" whether all projective sets are measurable and have the perfect set property. Luzin (1930), p. 323, elaborated:

> ...the author considers the question of knowing whether all projective sets are measurable to be insoluble, because in his view the definitions of projective set and Lebesgue measurability are incomparable and hence not logically related.

Luzin's prophecy was close to the mark, as later developments showed. The ZF axioms do not decide whether all the projective sets are measurable, or whether they all have the perfect set property—and one gets different answers depending on which additional axioms are used.

If one takes the axiom that all sets are constructible, as Gödel did in 1938 when proving the consistency of the axiom of choice (see Section 2.9), then one gets a definable well-ordering of $\mathbb{R}$, and with it a definable nonmeasurable set. Gödel noticed (apparently with some help from Ulam) that the definition puts the nonmeasurable set in the second level of the projective hierarchy. Moreover, one gets a CA set without the perfect set property. On the other hand, as we saw in Section 7.3, assuming the existence of infinitely many Woodin cardinals allows us to prove that all projective sets are determined, and hence measurable. The same assumption implies that all projective sets have the perfect set property, because the perfect set property is another of the nice consequences of determinacy.

Current thinking in set theory is that "inclusive" axioms, such as the large cardinal axioms, are more plausible than "restrictive" axioms such as Gödel's axiom of constructibility. It seems right, for example, that sets obtained by the axiom of choice should be as far from "definable" as possible, and large cardinal axioms seem to achieve this. But the larger the cardinal, the further we are from provability. Perhaps we have already crossed the line into contradiction. One does not know, and one may never know.

DO AXIOMS OF INFINITY SETTLE EVERYTHING?

It is not impossible that ... some completeness theorem would hold which would say that every proposition expressible in set theory is decidable from the present axioms plus some true assertion about the largeness of the universe of sets.

—Kurt Gödel
Remarks before the Princeton bicentennial conference, 1946,
in Gödel (1990), p. 151.

Soon after he proved his incompleteness theorems in 1930, Gödel began to think about the endless need for new and stronger axioms that his theorems imply. We cannot enumerate a complete sequence of such axioms in any computable way, but perhaps it is possible to say how increasing "strength" of axioms might be described mathematically. Set theory looks like a good arena for testing Gödel's "program," because it is good at measuring endless complexity—by ordinals and cardinals—and it is rich in unproved theorems. The history of large cardinals seems to bear this out, with Solovay, Martin, Steel, and Woodin using larger and larger cardinals to derive stronger and stronger theorems about sets of real numbers.

Indeed, it makes sense that if κ is a cardinal so large that that V_κ is a model of ZF and $\lambda > \kappa$ is a cardinal such that V_λ is also a model of ZF, then *the existence of λ implies everything and* more *than is implied by the existence of κ.* This is because V_κ is a model of ZF + "κ does not exist" (since κ is not in V_κ), whereas V_λ is a model of ZF + "κ exists" (since $\kappa < \lambda$ and hence κ is in V_λ). Thus one might begin to hope that any sentence not settled by the ZF axioms might be settled by a suitable large cardinal axiom.

Gödel was also aware that axioms of infinity decide sentences about *integers* that are unprovable in PA. For example, Con(PA) is not provable in PA and hence not in ZF − Infinity either, but in ZF we can prove Con(PA) by exhibiting a model for the PA axioms, such as V_ω. Thus the ordinary ZF axiom of infinity is strong enough to prove Con(PA). Stronger axioms of infinity can prove more. For example, Con(ZF) is not provable in ZF, by Gödel's second theorem, but it is provable in ZF + Inaccessible because, for inaccessible κ, V_κ is a model of ZF, and so on.

As mentioned in the previous section, large cardinal axioms can prove new theorems about integers, such as Con(ZF), beyond those provable from "small" cardinal axioms such as the ZF axiom of infinity. The new theorems include some of the form "$p(x_1, x_2, \ldots, x_n) = 0$ has no integer solution," where p is a polynomial with integer coefficients. This follows from the Matiyasevich work on polynomial equations mentioned in

Section 3.8. Thus the long arm of large cardinals reaches all the way down to basic arithmetic.

We also know that certain "miniature" axioms of infinity have an influence on PA: ε_0-induction proves Con(PA) and Goodstein's theorem, while Γ-induction proves Friedman's finite form of Kruskal's theorem, where Γ is a particular countable ordinal much larger than ε_0. Here ε_0-induction is analogous to the ordinary ZF axiom of infinity, and Γ-induction is perhaps the analogue of some axiom of inaccessibles. The theory of induction axioms and their consequences is not as developed as the theory of large cardinal axioms and their consequences, but we can expect more from it in the future. There should be not just two, but three, four, many roads to infinity.

Nevertheless, for all the successes of axioms of infinity, we must admit that they have thus far failed to settle the biggest question of set theory: the continuum hypothesis. In 1947, Gödel wrote a paper called *What is Cantor's Continuum Problem?* that expressed his hopes for axioms of infinity in a form similar to the quote above, but tempered with doubts about the axioms then known (see Gödel (1990), p. 182).

> As for the continuum problem, there is little hope of solving it by means of those axioms of infinity which can be set up on the basis of principles known today ... But probably there exist others based on hitherto unknown principles ...

In the 1960s, hopes were temporarily raised by the discovery of Hanf and Tarski that measurable cardinals are stronger than mere inaccessibles, and the subsequent discovery of many stronger cardinals. But by 1967 it was clear (from a result of Levy and Solovay) that Cohen's methods could be used to build models of set theory in which the continuum hypothesis was not affected by the new large cardinals.

Today, we know more axioms of infinity, but they still do not appear to settle the continuum hypothesis. On the other hand, neither do any other reasonable axioms, as far as we know. We can only hope, with Hilbert, that "we must know, we shall know."

BIBLIOGRAPHY

Irving H. Annellis. From semantic tableaux to Smullyan trees: a history of the development of the falsification tree method. *Modern Logic*, 1:36–69, 1990.

Emil Artin. Theorie der Zöpfe. *Abhandlungen aus dem Mathematischen Seminar der Universität Hamburg*, 4:42–72, 1925.

Jeremy Avigad. Number theory and elementary arithmetic. *Philosophia Mathematica*, 11(3):257–284, 2003.

Jeremy Avigad. The metamathematics of ergodic theory. *Annals of Pure and Applied Logic*, 157:64–76, 2009.

Reinhold Baer. Zur Axiomatik der Kardinalarithmetik. *Mathematische Zeitschrift*, 29:381–396, 1928.

Stefan Banach and Alfred Tarski. Sur la décomposition des ensembles de points en parties respectivement congruents. *Fundamenta Mathematicae*, 6:244–277, 1924.

Jon Barwise. *Handbook of Mathematical Logic*. North-Holland Publishing Co., Amsterdam, 1977. Edited by Jon Barwise, With the cooperation of H. J. Keisler, K. Kunen, Y. N. Moschovakis and A. S. Troelstra.

Emile Borel. *Leçons sur la théorie des fonctions*. Gauthier-Villars, Paris, 1898.

P. W. Bridgman. *Reflections of a Physicist*. Philosophical Library, New York, 1955.

Peter J. Cameron. *Combinatorics: Topics, Techniques, Algorithms*. Cambridge University Press, Cambridge, 1994.

Georg Cantor. Über unendliche, lineare Punktmannigfaltigkeiten. *Mathematische Annalen*, 21:545–591, 1883. English translation by William Ewald in Ewald (1996), volume II, pp. 881–920.

Alonzo Church. *Introduction to Mathematical Logic. Vol. I*. Princeton University Press, Princeton, N. J., 1956.

E. A. Cichon. A short proof of two recently discovered independence results using recursion theoretic methods. *Proceedings of the American Mathematical Society*, 87 (4):704–706, 1983.

Paul J. Cohen. The independence of the continuum hypothesis. *Proceedings of the National Academy of Sciences*, 50:1143–1148, 1963.

Paul J. Cohen. *Set Theory and the Continuum Hypothesis*. W. A. Benjamin, Inc., New York-Amsterdam, 1966.

Stephen A. Cook. The complexity of theorem-proving procedures. *Proceedings of the 3rd Annual ACM Symposium on the Theory of Computing*, pages 151–158, 1971. Association of Computing Machinery, New York.

Martin Davis, editor. *The Undecidable*. Dover Publications Inc., Mineola, NY, 2004. Corrected reprint of the 1965 original [Raven Press, Hewlett, NY].

Liesbeth De Mol. Tag systems and Collatz-like functions. *Theoretical Computer Science*, 390(1):92–101, 2008. doi: 10.1016/j.tcs.2007.10.020. URL http://dx.doi.org/10.1016/j.tcs.2007.10.020.

Richard Dedekind. *Essays on the Theory of Numbers*. Open Court, Chicago, 1901. Translated by Wooster Woodruff Beman.

Patrick Dehornoy. From large cardinals to braids via distributive algebra. *Journal of Knot Theory and its Ramifications*, 4(1):33–79, 1995.

Patrick Dehornoy. *Braids and Self-distributivity*. Birkhäuser Verlag, Basel, 2000.

Reinhard Diestel. *Graph Theory*. Springer-Verlag, Berlin, third edition, 2005.

William Ewald. *From Kant to Hilbert: A Source Book in the Foundations of Mathematics. Vol. I, II*. The Clarendon Press Oxford University Press, New York, 1996.

Dov M. Gabbay and John Woods. *Handbook of the History of Logic, volume 5*. Elsevier, Amsterdam, 2009. Edited by Dov M. Gabbay and John Woods.

George Gamow. *One, Two, Three, . . . , Infinity*. Viking Press, New York, 1947.

Gerhard Gentzen. Untersuchungen über das logische Schliessen. *Mathematische Zeitschrift*, 19:176–210, 1935. English translation by M.E. Szabo in Gentzen (1969), pp. 68–131.

Gerhard Gentzen. Die Widerspruchsfreiheit der reinen Zahlentheorie. *Mathematische Annalen*, 112:493–565, 1936. English translation by M.E. Szabo in Gentzen (1969), pp. 132–213.

Gerhard Gentzen. Beweisbarkeit und Unbeweisbarkeit von Anfangsfällen der transfiniten Induktion in der reinen Zahlentheorie. *Mathematische Annalen*, 119: 140–161, 1943. English translation by M.E. Szabo in Gentzen (1969), pp. 287–308.

Gerhard Gentzen. *The Collected Papers of Gerhard Gentzen*. Edited by M. E. Szabo. North-Holland Publishing Co., Amsterdam, 1969.

Jean-Yves Girard. *Proof Theory and Logical Complexity*. Bibliopolis, Naples, 1987.

Henry H. Glover, John P. Huneke, and Chin San Wang. 103 graphs that are irreducible for the projective plane. *Journal of Combinatorial Theory, Series B*, 27 (3):332–370, 1979.

Kurt Gödel. Über die Länge von Beweisen. *Ergebnisse eines mathematischen Kolloquiums*, 7:23–24, 1936. English translation in Gödel (1986), pp. 397–399.

Kurt Gödel. The consistency of the axiom of choice and the generalized continuum-hypothesis. *Proceedings of the National Academy of Sciences*, 24:556–557, 1938.

Kurt Gödel. Consistency proof for the generalized continuum hypothesis. *Proceedings of the National Academy of Sciences*, 25:220–224, 1939.

Kurt Gödel. *Collected works. Vol. I*. The Clarendon Press Oxford University Press, New York, 1986. Publications 1929–1936, Edited and with a preface by Solomon Feferman.

Kurt Gödel. *Collected Works. Vol. II*. The Clarendon Press Oxford University Press, New York, 1990. Publications 1938–1974, Edited and with a preface by Solomon Feferman.

Kurt Gödel. *Collected Works. Vol. V*. The Clarendon Press Oxford University Press, Oxford, 2003. Correspondence H–Z, Edited by Solomon Feferman, John W. Dawson, Jr., Warren Goldfarb, Charles Parsons and Wilfried Sieg.

R. L. Goodstein. On the restricted ordinal theorem. *The Journal of Symbolic Logic*, 9:33–41, 1944.

Ronald L. Graham, Bruce L. Rothschild, and Joel H. Spencer. *Ramsey Theory*. John Wiley & Sons Inc., New York, second edition, 1990.

Hermann Grassmann. *Lehrbuch der Arithmetik*. Enslin, Berlin, 1861.

G. H. Hardy. A theorem concerning the infinite cardinal numbers. *The Quarterly Journal of Pure and Applied Mathematics*, 35:87–94, 1904.

Felix Hausdorff. Grundzüge einer Theorie der geordneten Mengen. *Mathematische Annalen*, 65:435–505, 1908. English translation in Hausdorff (2005), pp. 197–258.

Felix Hausdorff. *Set Theory*. Chelsea Publishing Company, New York, 1937/1957. Translation of the 1937 German edition by John R. Aumann, et al.

Felix Hausdorff. *Hausdorff on Ordered Sets*. American Mathematical Society, Providence, RI, 2005. Translated from the German, edited and with commentary by J. M. Plotkin.

David Hilbert. Neubegründung der mathematik (erste mitteilung). *Abhandlungen aus dem mathematischen Seminar der Hamburgischen Universität*, 1:157–177, 1922. English translation in Ewald (1996), vol. II, pp. 1117–1134.

David Hilbert. Über das Unendliche. *Mathematische Annalen*, 95:161–190, 1926. English translation in van Heijenoort (1967), pp. 367–392.

Thomas Jech. *Set Theory*. Springer-Verlag, Berlin, third edition, 2003.

Akihiro Kanamori. *The Higher Infinite*. Springer-Verlag, Berlin, second edition, 2003.

Akihiro Kanamori and Menachem Magidor. The evolution of large cardinal axioms in set theory. In *Higher Set Theory*, pages 99–275. Springer, Berlin, 1978.

Laurie Kirby and Jeff Paris. Accessible independence results for Peano arithmetic. *The Bulletin of the London Mathematical Society*, 14(4):285–293, 1982.

Dénes König. *Theory of Finite and Infinite Graphs*. Birkhäuser Boston Inc., Boston, MA, 1936/1990. Translated from the 1936 German edition by Richard McCoart.

Kazimierz Kuratowski. Sur l'état actuel de l'axiomatique de la théorie des ensembles. *Annales de la Société Polonaise de Mathématique*, 3:146–149, 1924. In Kuratowski (1988), p. 179.

Kazimierz Kuratowski. Sur le problème des courbes gauches en topologie. *Fundamenta Mathematicae*, 15:271–283, 1930. In Kuratowski (1988), pp. 345–357.

Kazimierz Kuratowski. *A Half Century of Polish Mathematics*. Pergamon Press Inc., Elmsford, N.Y., 1980. Translated from the Polish by Andrzej Kirkor, With a preface by S. Ulam.

Kazimierz Kuratowski. *Selected Papers*. PWN—Polish Scientific Publishers, Warsaw, 1988.

Bruce M. Landman and Aaron Robertson. *Ramsey Theory on the Integers*. American Mathematical Society, Providence, RI, 2004.

Azriel Levy. *Basic Set Theory*. Springer-Verlag, Berlin, 1979.

Nikolai N. Luzin. Sur les ensembles projectifs de M. Henri Lebesgue. *Comptes Rendus de l'Academie des Sciences, Paris*, 180:1572–1574, 1925.

Nikolai N. Luzin. *Leçons sur les ensembles analytiques*. Gauthier-Villars, Paris, 1930. Reprinted with corrections by Chelsea, New York, 1972.

Donald A. Martin. Measurable cardinals and analytic games. *Fundamenta Mathematicae*, 66:287–291, 1969/1970.

Herman Melville. Hawthorne and his mosses. *Literary World*, 24, August 17, 1850. In *Herman Melville. Pierre, Israel Potter, The Piazza Tales, The Confidence Man, Uncollected Prose, and Billy Budd*, Library of America, 1984, pp. 1154–1171.

Elliott Mendelson. *Introduction to Mathematical Logic*. D. Van Nostrand Co., Inc., Princeton, NJ, 1964.

Gregory H. Moore. *Zermelo's Axiom of Choice*. Springer-Verlag, New York, 1982.

Theodore S. Motzkin. Cooperative classes of finite sets in one and more dimensions. *Journal of Combinatorial Theory*, 3:244–251, 1967.

Reviel Netz and William Noel. *The Archimedes Codex*. Weidenfeld and Nicholson, London, 2007.

Jeff Paris and Leo Harrington. A mathematical incompleteness in Peano arithmetic. 1977. In Barwise (1977), pp. 1133–1142.

Emil L. Post. Absolutely unsolvable problems and relatively undecidable propositions—an account of an anticipation, 1941. In Davis (2004), pp. 338–433.

Emil L. Post. Formal reductions of the general combinatorial decision problem. *American Journal of Mathematics*, 65:197–215, 1943.

Emil L. Post. *Solvability, Provability, Definability: the Collected Works of Emil L. Post*. Birkhäuser Boston Inc., Boston, MA, 1994. Edited and with an introduction by Martin Davis.

Frank P. Ramsey. On a problem of formal logic. *Proceedings of the London Mathematical Society*, 30:264–286, 1930.

Gerhard Ringel. *Map Color Theorem*. Springer-Verlag, New York, 1974.

Paul C. Rosenbloom. *The Elements of Mathematical Logic*. Dover Publications Inc., New York, N. Y., 1950.

Stephen G. Simpson. *Subsystems of Second Order Arithmetic*. Springer-Verlag, Berlin, 1999.

Raymond M. Smullyan. *Theory of Formal Systems*. Annals of Mathematics Studies, No. 47. Princeton University Press, Princeton, NJ, 1961.

E. Szemerédi. On sets of integers containing no k elements in arithmetic progression. *Acta Arithmetica*, 27:199–245, 1975.

Terence Tao. *Structure and Randomness: Pages from Year One of a Mathematical Blog*. American Mathematical Society, 2009.

A. S. Troelstra and H. Schwichtenberg. *Basic Proof Theory*. Cambridge University Press, Cambridge, 1996.

A. M. Turing. On computable numbers, with an application to the Entscheidungsproblem. *Proceedings of the London Mathematical Society*, 42:230–265, 1936.

Alasdair Urquhart. Emil Post. 2009. In Gabbay and Woods (2009), pp. 617–666.

Mark van Atten and Juliette Kennedy. Gödel's Logic. 2009. In Gabbay and Woods (2009), pp. 449–510.

Jean van Heijenoort. *From Frege to Gödel. A Source Book in Mathematical Logic, 1879–1931.* Harvard University Press, Cambridge, Mass., 1967.

Stan Wagon. *The Banach-Tarski Paradox.* Cambridge University Press, Cambridge, 1993.

S. S. Wainer. A classification of the ordinal recursive functions. *Archiv für Mathematische Logik und Grundlagenforschung,* 13:136–153, 1970.

Alfred North Whitehead and Bertrand Russell. *Principia Mathematica,* volume 1. Cambridge University Press, Cambridge, 1910.

Robert S. Wolf. *A Tour through Mathematical Logic.* Mathematical Association of America, Washington, DC, 2005.

Ernst Zermelo. Beweis dass jede Menge wohlgeordnet werden kann. *Mathematische Annalen,* 59:514–516, 1904. English translation in van Heijenoort (1967), pp. 139–141.

Ernst Zermelo. Untersuchungen über die Grundlagen der Mengenlehre I. *Mathematische Annalen,* 65:261–281, 1908. English translation in van Heijenoort (1967), pp. 200–215.

Ernst Zermelo. Über Grenzzahlen und Mengenbereiche. *Fundamenta Mathematicae,* 16:29–47, 1930.

INDEX

and Heine-Borel, 116
  in Ramsey theory, 152
König, Dénes, 116, 158
Kreisel, Georg, 138
Kruskal, Joseph, 155
  theorem, 155, 157
Kuratowski, Kazimiercz
  and inaccessibles, 177
  greatest discovery, 178
  on Lebesgue's mistake, 178
  theorem, 159
    Kuratowski version, 160
    Wagner version, 160

least upper bound
  of countable ordinals, 35
  of ordinals, 29, 31, 43
  of reals, 12, 22
  uniqueness, 44
Lebesgue, Henri, 56, 62
  and axiom of choice, 61
  and Baire hierarchy, 58
  and Borel hierarchy, 59, 179
  measure, 60
  mistake, 178
    discovered by Suslin, 179
Leibniz, Gottfried Wilhelm, 89
less than relation, 27
Levin, Leonid, 115
Levy, Azriel, 24, 182
limit
  of an infinite process, 20
  ordinal, 35, 38, 43, 58
  process in analysis, 56
Lindemann, Ferdinand, 11
linear ordering, 33, 120
  of braids, 173, 175
  of Cantor normal forms, 142
Liouville, Joseph, 10
  transcendental number, 10
listing, 1
  and counting, 1, 2
  equations, 11
  functions, 15
  sets, 6, 74
  strings, 36, 63, 72, 73
  theorems, 73

logic
  axioms, 90
  completeness of, 97
  first order, 107
  predicate see predicate logic 97
  propositional see propositional
    logic 97
  second order, 107, 136
    lacks compactness, 136
    lacks completeness, 136
Luzin, Nikolai N., 8
  and analytic sets, 179
  and descriptive set theory, 179
  and projective sets, 179
  on the line, 8
  prophecy about projective sets,
    180

Mahlo, Paul, 177
Markov, Andrei A., 92, 93
Martin, Donald A., 171, 176, 181
Matiyasevich, Yuri V., 93
  and polynomial equations, 181
Mazur, Stanisław, 170
measurability
  and inaccessible cardinals, 169,
    182
  Borel, 60
  implies inaccesibility, 178
  Lebesgue, 60, 169
  of analytic sets, 179
  of cardinal numbers, 170
  of determined sets, 171
  of projective sets, 180
  relative consistency, 169, 178
measure, 12
  Borel, 60
  countable additivity property, 58,
    61
  difference property, 58
  finitely additive, 61
  for all sets of reals, 62, 64
  Lebesgue, 60
  of a set of reals, 58
  of countable set, 13
  theory, 60

Printed in the United States
by Baker & Taylor Publisher Services